MEMORIAL UNIVERSITY
SEP 13 2001
LIBRARY
OF NEWFOUNDLAND
WITHDRAWN

D1416041

TEAM PERFORMANCE MANAGEMENT

Advances in Interdisciplinary Studies of Work Teams

Michael M. Beyerlein, Douglas A. Johnson, and Susan T. Beyerlein, Series Editors

Volume 1: *Theories of Self-Managing Work Teams*
edited by Michael M. Beyerlein and Douglas A. Johnson, 1994

Volume 2: *Knowledge Work in Teams*
edited by Michael M. Beyerlein, Douglas A. Johnson, and Susan T. Beyerlein, 1995

Volume 3: *Team Leadership*
edited by Michael M. Beyerlein, Douglas A. Johnson, and Susan T. Beyerlein, 1996

Volume 4: *Team Implementation Issues*
edited by Michael M. Beyerlein, Douglas A. Johnson, and Susan T. Beyerlein, 1997

Volume 5: *Product Development Teams*
edited by Michael M. Beyerlein, Douglas A. Johnson, and Susan T. Beyerlein, 2000

TEAM PERFORMANCE MANAGEMENT

Edited by

MICHAEL M. BEYERLEIN
Center for the Study of Work Teams
Department of Psychology
University of North Texas

DOUGLAS A. JOHNSON
Center for the Study of Work Teams
Department of Psychology
University of North Texas

SUSAN T. BEYERLEIN
Center for the Study of Work Teams
Department of Psychology
University of North Texas

JAI PRESS INC.
Stamford, Connecticut

Library of Congress Cataloging-in-Publication Data

Team performance management / edited by Michael M. Beyerlein, Douglas A. Johnson, Susan T. Beyerlein.

p. cm. -- (Advances in interdisciplinary studies of work teams ; vol. 6)

"This book consists of chapters based on presentations at the University of North Texas sixth annual Symposium on Work Teams"--Introd.

Includes bibliographical references.

ISBN 0-7623-0655-6 (alk. paper)

1. Teams in the workplace--Congresses. I. Beyerlein, Michael Martin. II. Johnson, Douglas A. III. Beyerlein, Susan T. IV. Series.

HD66. T425 2000
657.4'03--dc21

00-032261

100 Prospect Street
Stamford, Connecticut 06901-1640

ISBN: 0-7623-0655-6

Library of Congress Card Catalog Number 00-032261

Manufactured in the United States of America

CONTENTS

LIST OF CONTRIBUTORS

Vikas Anand	Sam Walton College of Business University of Arkansas
Donde P. Ashmos	Division of Management and Marketing University of Texas at San Antonio
Michael Barriere	Industrial Organizational Psychology Hofstra University
Elizabeth Blickensderfer	Naval Air Warfare Center Training Systems Division
Janis A. Cannon-Bowers	Naval Air Warfare Center Training Systems Division
L. Scott Casino	Department of Management Pamplin College of Business, Virginia Tech
Nancy J. Cooke	Department of Psychology New Mexico State University
Robert E. Debold, Jr.	The Center for Fourth Wave Concepts
Dennis Duchon	Division of Management and Marketing University of Texas at San Antonio
Lillian T. Eby	Department of Psychology University of Georgia
Mel Fugate	Department of Management Arizona State University

Ira T. Kaplan Industrial Organizational Psychology
Hofstra University

Charles C. Manz Department of Management
University of Massachusetts, Amherst

William Metlay Industrial Organizational Psychology
Hofstra University

Dana M. Milanovich Naval Air Warfare Center
Training Systems Division

Maria Nathan Division of Management and Marketing
University of Texas at San Antonio

Christopher P. Neck Department of Management
Pamplin College of Business, Virginia Tech

Anthony G. Parisi Department of Psychology
University of Georgia

Eduardo Salas Department of Psychology
University of Central Florida

Paul F. Skilton Department of Management
Grand Valley State University

Wanda J. Smith Department of Management
Pamplin College of Business, Virginia Tech

Vicki Smith-Daniels Department of Management
Arizona State University

Kimberly A. Smith-Jentsch Naval Air Warfare Center
Training Systems Division

Renée Stout Naval Air Warfare Center
Training Systems Division

Duane Windsor | Jesse H. Jones Graduate School of Management
Rice University

INTRODUCTION

This volume consists of chapters based on presentations at the University of North Texas' sixth annual Symposium on Work Teams. The chapters provide in-depth coverage of key topics by some of the world's leading thinkers on team development and address the "how's" and "why's" that underlie team effectiveness, including such issues as learning, compensation, and empowerment.

Team development is an issue of intense interest to people in the work place. Work teams are being tried in many settings, including manufacturing, research and development (R&D), support services, government, education, sales, and so on. In some of those places, the effort is abandoned for a variety of reasons, including changes in management personnel, business pressures focused on the short term, the size of the investment in developing teams, and the slow pace of team development. High performance levels do not automatically flow from the implementation of teams; they flow from the proper design and continued support of the teams, including empowerment, skill building, reward system changes, and information access.

The chapters in this volume address such issues of design and support. Questions such as the following are addressed:

- How should knowledge be managed to enhance empowerment?
- Can revolutionary methods of measurement track performance in complementary ways?

- How does sharing mental models improve team decision making?
- What contingencies should reward systems incorporate?
- How do technical and social skills fit together?
- How can performance management be adapted to teams?
- What issues of team and organization design emerge with the recognition that companies are complex adaptive systems responding to dynamic environments?

Some of the highlights of each chapter are described below in the order they appear in the book.

Empowerment comes in two forms: decision-making authority and increased resources and abilities. The two may be related. Knowledge has become an increasingly obvious enabler of teams. Access to information and the skill to use it have become critical for empowered team performance in an environment with increased change, globalization and competition. Vikas Anand, Charles Manz, and Mel Fugate build a model of team memory that points the way to breaking through the limits of traditional knowledge management by teams. The team has three problems to solve: accessing knowledge held by outsiders, increasing the proportion of obtained knowledge that is actually utilized, and communicating tacit knowledge. The chapter suggests strategies for dealing with each problem.

In their chapter on team performance measurement, Renée Stout, Nancy Cooke, Eduardo Salas, Dana Milanovich, and Janis Cannon-Bowers build on a sequence of careful studies about the knowledge, skills, and attitudes necessary for team effectiveness. The authors focus on how team performance measures can be enhanced by the use of cognitive engineering techniques, including observation, interviews, process tracing, and concept mapping using Pathfinder software that can access more tacit knowledge. Based on a thorough review of team performance and its measurement, a discussion of cognitive engineering reveals the underlying cognitive components of performance. Examples include adaptability, coordinated action and decision making based on such knowledge competences as understanding of team-member roles, cue-strategy associations, and shared mental models. The examples were studied in the field in their natural settings to reveal the perceptual, attentional, memory, and decision-making structures and processes underlying complex task performance.

Learning from experience requires deliberate reflection. A well-known example is the comparison of the "person with 30 years of experience with the person with one year of experience 30 times." The former learned continuously; the latter did not. Teams may replicate this example. Teams that learn continuously need a discipline that is regularly practiced. Kimberly Smith-Jentsch, Elizabeth Blickensderfer, Eduardo Salas, and Janis Cannon-Bowers propose that guided team self-correction represents that discipline. It rests on the team developing shared mental models during structured debriefings. In an environment with training, time for meeting, a facilitator, and some skill development, a team ought to generate significant insights through these disciplined debriefings that will directly impact performance.

Self-managing work teams (SMWTs or SMTs) are small social and technical systems with members that have joint responsibilities for outcomes. The ability of the team to reach its goals depends on the abilities of each member and the skill they have in working together. Training, formally and informally, should increase those competencies. However, who decides which members should get which training? Wanda Smith, Scott Casino, and Christopher Neck examine how individual members influence the rest of the team around training decisions. The style of influence varies depending on whether the training is intended to build social skills or technical skills for the member. The styles range from use of coercion to use of cooperation tactics depending on the climate of the team. If a team is to make its own training decisions, awareness of the dynamics of influence may improve that decision-making process.

New product development teams have special learning challenges to overcome. The teams consist of people from multiple disciplines and with a wide range of responsibilities. Integration of such diverse membership into a high performing team takes special skills. Paul Skilton and Vicki Smith-Daniels suggest that those skills include: (1) being able to translate between different areas, (2) knowing who knows what/where to find information, and (3) the ability to initiate and sustain dialogue, including managing conflict, getting feedback, and generating consensus. Skilton and Smith-Daniels refer to their skills model as a T-shaped model. The integration skills overlay the deep functional expertise that the members bring from their disciplines, similar to the way the bar lies on top of the T. The authors suggest that the T-skills approach develops capability to the point that a team can accomplish things that a lone genius cannot.

Michael Barriere, Ira Kaplan and William Metlay focus on the use of behavior modification to manage team performance. Whether referred to as organizational behavior modification, organizational behavior management, or performance management, this approach uses a systematic, data-based approach to link positive reinforcement to performance. The approach typically focuses on performance by individuals which may suboptimize the team's performance. To offer an alternative for enhancing team performance, the authors present a framework based on systems theory. The framework provides a matrix of 16 cells aligning Task, Individual, Group, and the group's Environment with Input, Process, Output, and Feedback. The authors then review carefully selected studies to determine which facets of the framework have been tested in prior studies. Several studies assessed group level facets, with the most comprehensive addressing 10 of the 16 cells. Several studies concluded that focusing reinforcement at the individual level failed to maximize performance at the team level.

Work teams are more complex systems than traditional work groups—empowered group action in contrast to supervised individual activity; consequently, they require a larger investment in training and other support activities. Are they worth that investment? Or, rather the question should be: when are they worth that extra investment? Duane Windsor writes about team evaluation and compensation which are post-performance activities. However, the design of the work and the specification of the team's work objectives should link to those activities. The decisions about evaluation and compensation involve comparisons of team performance with team objectives, and to the likely level of performance by conventional work groups. Measurement and evaluation require specification of task, outcome, action, process, time horizon, resources, constraints, and efficiencies. Evaluation can then follow two tracks: efficiency of resource utilization or a combination of process and outcomes.

There are least a dozen models of group development in the research and consulting literatures. The models agree on several things—development is necessary, it occurs across time, it is not guaranteed. The models all imply that the group is transformed into a different type of system in each new stage. This suggests that the way a team is handled by the organization ought to change also—something of a contingency theory of group development. Anthony Parisi and Lillian Eby present a contingency theory of rewards. Combining the literature from group development, time and work pacing, and equity versus equality norms, they create a framework that guides changes in reward allocation at critical points in a project. They argue that team members' preferences change at those points, thus justifying a shift from equity-based rewards to equality-based rewards.

A number of experts suggest that society is going through a revolution with changes as dramatic as those of the Industrial Revolution in the nineteenth century. The growing dependence on knowledge creation, acquisition, and utilization in organizations and their networks is creating a shift in perspective that requires a new vocabulary of concepts. Robert Debold describes the Fourth Wave Society (the knowledge wave) and argues that it enables us to build "holonic performance management systems." These self-organizing systems depend on feedback from the environment to be optimally adaptive. Examples of such organizations are beginning to appear in high tech companies around the world. Aspects of the TCG software consortium in Australia, Acer Computer in Taiwan, ABB in Europe (and the rest of the world), and Gore, which produces gore-tex fabric in the U.S., embody some of the principles Debold presents. These companies are exemplary in their adaptability, flexibility, and ability to take care of their employees.

Traditional views of organizing are linear; they argue for simple structures with management control of resources and decisions. In stable environments a linear approach may be adequate. Environments even in historically stable industries like banking are no longer stable; deregulation, globalization, competition, and technical change have created turbulence that demands adaptation. Traditional managers still instinctively fail to recognize how environmental turbulence drives change in their organizations. Dennis Duchon, Donde Ashmos, and Maria Nathan argue for a view of organizations as complex adaptive systems that generate adaptive ideas. Management must permit and support such spontaneous self-organization or emergence. The role of leadership has changed. And, a new role has emerged for teams—sensemaking. When the support, opportunity, and competencies are in place, the team can contribute to making sense in a dynamic environment that will continue to seriously challenge organizations using traditional approaches.

Each of these chapters makes a valuable contribution to our understanding of the effective development of teams. Each contains useful ideas for both practice and theory building. Taken as a whole, this volume should help serious thinkers in both business and academia improve their models for creating and sustaining effective work teams.

Michael M. Beyerlein
Douglas A. Johnson
Susan T. Beyerlein
Editors

FOREWORD

Lately I have found myself challenged to follow the patterns of change that have been flowing around and through our organizations and the world at large. Perhaps that is why one of the current organizational theories, complexity theory, is so aptly named. While there are patterns, they are not linear and not easily ascertained, certainly not with the naked eye.

In scanning the world of organization theory and practice, however, there do appear to be some trends that have been in place and are continuing. One of these is teaming. While the idea that groups of people could come together in a cooperative manner and become to some degree empowered has been around for at least 50 years. However, at first this notion was slow to evolve in practice, having to develop in organizations bred out of the industrial age. There is no doubt that the advent of the information age has accelerated the trend toward teaming.

I would venture the prediction that, as globalization continues, it will be another force that supports the teaming trend. Successful global ventures stretch the need for teamwork beyond even current practice. The need for collaboration across cultural, national, and geographic boundaries will drive us to better develop and use approaches to building teamwork.

Because I believe that the need for knowledge about how to achieve teamwork will continue to grow, I find it most gratifying to see the continued contributions to this field by the Center for Study of Work Teams. The chapters in this volume are drawn from the Center's most recent conference, one which brings practitioners and researchers together to share knowledge and experience. The chapters cover a diversity of issues and variables that contribute to successful teaming. They also cover a variety of types of teams and remind us that different factors apply to different team requirements. Collectively these chapters expand our knowledge base in the field.

As someone who has been both a researcher and practitioner, I can say that this volume is an example of how the Center has contributed to the accumulation and management of knowledge about teaming over an extended period of time. Anyone who is involved with teaming, either at the micro or macro level, will benefit from the contents of this volume and from the overall body of work it represents.

Laurie Broedling
President
LB Organizational Consulting

ACKNOWLEDGMENTS

The chapters in this volume grew out of presentations given at the University of North Texas' sixth annual Symposium on Work Teams. The presentations were given by the authors, but another key part of the Symposium consisted of the discussant remarks by representatives of business. Discussants typically talked with the authors and often read early drafts of their papers before the Symposium. Discussants were charged with sharing ways in which their companies were applying the concepts presented by the authors. The Center for the Study of Work Teams has considered the bridging of the gap between the academic and practice worlds, a mission to be pursued at each conference event. The success of that bridging is primarily due to the efforts of the discussants. Therefore, we acknowledge their important contribution to the Symposium and to the authors' thinking about the concepts in their papers from a practical frame of reference.

The discussants and their affiliations at that time of the Symposium were:

Kyle Ramsey, Intel Corporation
Richard Thier, Xerox Corporation
Dallen Miner, AMOCO Corporation
Tom Ruddy, Xerox Corporation
John Gilberti, First American Financial
Les Killingsworth, GTE Directories
William Sechrest, Winstead, Sechrest, & Minick

Leon Abbott, Lockheed Martin Tactical Aircraft Systems
Susan Gressett, VHA, Inc.
R. D. Ryza, Shell Oil Company

Finally, we want to acknowledge the help and support of Melanie Bullock. She has been responsible for communicating with authors, discussants, and editors, arranging for flow of manuscripts back and forth among these people, educating all of us on manuscript format, maintaining the relationship with the publisher, arranging for proofing, catching the errors the editors miss and bringing the chapters, introduction, and preface together in a final assembly to complete a whole for the publisher. Every year she has taken on more responsibility and leadership on the volumes in this series. In addition, she organized and oversaw all of the logistics of the Symposium event. It would not be inaccurate to think of her as project manager on this volume!

EMPOWERING WORK TEAMS WITH KNOWLEDGE

ENHANCING PERFORMANCE IN SERVICE AND INFORMATION-ORIENTED INDUSTRIES

Vikas Anand, Mel Fugate, and Charles C. Manz

ABSTRACT

In this chapter we make the case that team knowledge is an increasingly critical aspect of organizational performance and competitiveness, therefore, to manage team knowledge is to manage team performance. To better understand the construct of team knowledge we develop a model of team memory that provides insight into the effective acquisition and utilization of knowledge in the team context. We describe numerous organizational activities—socialization tactics, training of insiders and outsiders, the implementation of appropriate communications technology, utilization of teams approach—that can enhance both potential team knowledge and realized team knowledge, which collectively serve to empower teams with knowledge.

Advances in Interdisciplinary Studies of Work Teams, Volume 6, pages 1-26.

ISBN: 0-7623-0655-6

INTRODUCTION

It has been said that "knowledge is power." In this chapter we will explore the various ways in which organizations can ensure that their work teams are provided with relevant knowledge. We will rely on group and organizational memory perspectives to address team knowledge empowerment and to generate a series of propositions that can help guide future research into this important issue for contemporary organizations.

A primary advantage of using teams in organizations is that members can collectively apply larger amounts of knowledge to critical activities (Lawler, 1992). Not only are the skills and experiences of individual team members brought to bear on the tasks of the group, but the combinations of these assets also make available powerful synergies not available in a non-team context. However, increasing complexity and turbulence in business environments, coupled with the spread of globalization has greatly increased the amount of knowledge required to successfully perform key organizational tasks (Anand, Manz, & Glick, 1998; D'Aveni, 1994; Huber & Glick, 1993). Knowledge is now recognized as a key to building competitive advantage (Grant, 1996). Since most knowledge-intensive tasks are performed in teams (Laudon & Starbuck, 1996), an organization's ability to provide its work teams with relevant knowledge can provide it with a strong competitive edge.

This chapter focuses on the means by which teams in an organization are provided with relevant knowledge and information that allow them to effectively perform their jobs. We are primarily concerned with teams that perform knowledge and information intensive tasks such as strategy formulation, product development, research and development (R&D), software development, and so forth. However, many of our arguments will also generalize to most work teams. Further, since the proportion of knowledge-intensive work in organizational activities is continuously increasing in developed economies (Laudon & Starbuck, 1996), our arguments have wide theoretical and practical implications.

In this chapter we make the argument that the knowledge required by a team is generally in excess of the knowledge personally held by team members. We also argue that this gap is increasing with time. We use insights from a model of team memory to examine the ways in which organizations can reduce this gap. We argue that providing team members with additional knowledge is unlikely to close this gap. Rather, organizations will need to focus on a large number of support activities such as socialization, use of information technology, training of key outsiders (suppliers, distributors, etc.), providing behavioral and communication training to team members, and allowing for flexible norms and structures to ensure that knowledge is available in their work teams. We also argue that the continuous use of a team approach initiates a self-reinforcing loop such that the longer a team approach has been used in an organization, the greater the knowledge that is available to members.

We propose a model of team memory and then show that this model may be used to understand the knowledge available to a work team. This approach to examining team knowledge helps clarify our understanding of team knowledge—a concept that has been dealt with in fuzzy terms in the past. Indeed, clarification of the team knowledge construct provides an important first step towards the measurement of team knowledge. We then use the insights provided from the model of team memory to logically generate propositions linking organizational activities to team knowledge. These propositions are intended to provide a guide for organizations using a team approach and attempting to compete in today's knowledge-based society, and are also intended to guide future empirical research into the area.

Our chapter begins with an examination of knowledge, its impact on teams and a discussion of trends that impact knowledge availability in teams. We then describe a model of group memory and adapt it to work teams in organizations. We use insights from the model to identify ways in which organizations can increase knowledge availability in teams. We conclude with directions for future research and implications for practice.

KNOWLEDGE AND WORK TEAMS

Previous literature has highlighted the need to provide work teams with adequate knowledge. Self-managing teams need to possess knowledge so that they can effectively redesign their work and processes in response to work-environment variations (Manz & Stewart, 1997; Susman, 1976). Lawler (1992) and Manz (1992) point out that team members need to be provided task- and organization-related information, such as organizational goals and priorities, so that they may more effectively interact with the organization's environment. Klein (1994) argued that team members must be provided expertise in a variety of overlapping functions to provide teams with required flexibility. Additionally, Manz and Sims (1993) suggest that as teams move toward a self-leading mode, team members need to be trained in self-management and interpersonal skills.

The above (and many other) researchers make two critical generalizations or assumptions. First, most authors treat knowledge as a given and do not differentiate between the very different natures of the knowledge they refer to. For instance, when Lawler talks about organizational goals and priorities, the focus is on information that is relatively easily communicated. When Jackson (1996) focuses on expertise, the focus is on noncommunicable knowledge or tacit knowledge (Polanyi, 1966). These differences in knowledge are important, because the processes by which different kinds of knowledge and information are acquired and used in a team are different (Daft & Weick, 1984; Nonaka, 1994). Later in this section, we address this limitation by clearly defining the term knowledge as used in this chapter.

A second major assumption relates to the deterministic stance taken by researchers with respect to a team's knowledge requirements. It is typically assumed that most knowledge required by a team is determinable in advance, and that this knowledge can easily be transferred to and among team members. We challenge this assumption because all knowledge requirements are not determinable in advance, and even if they were, many times it may be impossible to effectively transfer the required knowledge to the appropriate team member(s) (Anand & Manz, 1997).

Defining Knowledge

Data, information, and knowledge have been the subject of intense study by organizational researchers; as a result classifications abound (Anderson, 1983; Nonaka, 1994; Polanyi, 1966; Sackman, 1992). Following the lead of Daft and Huber (1987), we classify knowledge in terms of its inherent communicability. Accordingly, we divide knowledge into two components: information and soft knowledge. The former is comprised of all knowledge and data that can easily be encoded and shared among individuals. Information therefore encompasses data about the internal and external work environment, which includes explicit and procedural knowledge which can be written down.

Soft knowledge, on the other hand, refers to all noncommunicable forms of knowledge including forms such as tacit knowledge (e.g., the knowledge used by wine tasters to identify the quality of wine) that cannot be written down and shared among individuals. It also includes individual expertise and skills such as good judgment, intuitive abilities, ability to make sense of ambiguous and incomplete data, and learned behaviors such as self-management (Agor, 1991; Anand, Skilton, & Keats, 1996; Bazerman, 1997; Manz & Sims, 1980). Our description of soft knowledge is similar to the two components of tacit knowledge defined by Nonaka and Takeuchi (1995): cognitive and technical. The cognitive element consists of mental analogies and models held by an individual which describe to him or her how the world works. The technical elements consist of skills and expertise. Both of these components are generally acquired over time and experience through processes such as vicarious learning, apprenticeship, extended interaction, and repetition (Anand et al., 1998; Bandura, 1986; Nonaka, 1994; Nonaka & Takeuchi, 1995). Thus, in the remainder of the chapter we use the term *knowledge* to inclusively refer to both information and soft knowledge. When the focus is on only one of these two components, it is referred to specifically.

Team Knowledge Requirements Versus Team Knowledge Availability

Contemporary organizations are faced with a major knowledge management dilemma: the knowledge required to perform critical organizational activities is increasing due to continuous growth in the service sector, increasing turbulence and globalization of business environments, and increasing recognition that knowledge is a source of organizational competitive advantage (Conner & Prahalad, 1996; Spender, 1996). On the other hand, we point out that there are limits to the amount of knowledge that teams can possess. Thus, meeting the ever-increasing knowledge requirements has become a challenge for firms. Firms that can best meet the increased knowledge demands are likely to be superior performers.

Service industries include businesses such as software development, telecommunications, research and development, banking, insurance, and consulting. The contribution of service industries to the economy has been steadily increasing. For instance, in 1990, the service sector employed over half of U.S. workers and constituted about three fourths of the GNP (Laudon & Starbuck, 1996). This growth has been fueled by the increasing transfer of manufacturing activities to third world nations, and by increased automation of the work place that has significantly reduced domestic manufacturing jobs (Laudon & Starbuck, 1996). Since knowledge and information are resources more central to service sector organizations, the increase in knowledge requirements has paralleled the growth of this industry (Quinn, 1993).

Additionally, globalization and the growth of information technology has significantly increased the turbulence, unpredictability, and complexity in business environments (D'Aveni, 1994; Huber & Glick, 1993). This has affected knowledge requirements for organizational activities in several ways. First, as a larger number of factors need to be considered for performing activities—for example, legal compliance across global markets and cultural sensitivity in transnational initiatives—the amount of knowledge required has increased exponentially (Anand & Manz, 1997).

Second, the rapid and often dramatic changes in environmental conditions create the need for continuous learning in order to remain effective at existing or even new tasks. And, since the knowledge requirements change rapidly, even if team members possess appropriate knowledge at a given point in time, it soon becomes obsolete.

Third, as knowledge is clearly recognized as a primary determinant of organizational success, firms are increasingly competing on the basis of knowledge. This is reflected in the development of masters-level programs in information management (or related areas) by business schools, as well as the increasing frequency of Chief Information Officers (CIOs) in many large organizations (Caldwell, 1996; Machlis, 1997). However, when a firm increases the knowledge used in delivering its product to the market place, its competitors have no option but to follow suit. This places organizations in a competitive loop, continuously increasing the knowledge they must apply to their activities.

Clearly then, knowledge requirements are increasing at a torrid pace in many organizations. However, the knowledge held by team members can increase only to a more limited extent. A team's knowledge is primarily held in the memories of its members (Anand et al., 1998). Since the storage capacity of individual memories is limited, it follows that the amount of knowledge held by teams is also limited (Ashcraft, 1989; Fiske & Taylor, 1984). Further, increasing the knowledge available to a team by expanding its membership is not a viable option because team effectiveness declines if its membership exceeds an optimal level (Cummings, 1978). Additionally, while the advent of advanced information technologies has significantly improved the information processing abilities of individuals, such advances have mainly been restricted to information, and thus do not incorporate soft knowledge (Anand et al., 1998; Nunnamaker, Briggs, Mittleman, Vogel, & Balthazar, 1996). And since current business conditions also require significant amounts of soft knowledge, advanced information technologies are at best partial solutions (Agor, 1991; Anand et al., 1998). This suggests two challenges for organizations. Not only must they learn and continuously acquire knowledge relevant to their dynamic environments, they must also ensure that the relevant knowledge is transferred to their various work teams.

Finally, we are in an age where most industries operate in *hypercompetitive* environments (D'Aveni, 1994). Such environments are characterized by their rapid change and unpredictability. This creates an additional challenge for organizations. Because future environmental conditions are unknown, the knowledge required in those conditions is also unknown. Consequently, organizations can anticipate and proactively provide team members with only limited amounts of knowledge that will be required for their future needs.

The preceding discussion highlights the fact that team knowledge requirements are growing at a fast pace and that traditional organizational practices of providing knowledge to team members do not adequately meet team knowledge requirements. We consider a team to be knowledge empowered when it is in a position to access and utilize all knowledge relevant to tasks. Clearly, teams will differ in the degree to which they are knowledge empowered and successful organizations of the future will be those that can ensure a high degree of knowledge empowerment to their teams.

TEAM MEMORY

Knowledge used by a team is rarely acquired at the time of use. Rather, team members generally acquire knowledge well in advance, and retrieve it as needed. In between these two points in time, knowledge is stored. Since a team's knowledge undergoes the three basic memory processes—acquisition, storage, and retrieval—it makes sense to examine a team's knowledge using the metaphorical construct of *team memory* (Ashcraft, 1989).

Figure 1 describes a model of team memory. The *potential memory* of a team is comprised of the knowledge held by all of its members, and the knowledge that members can access from others, both inside and outside the organization. The

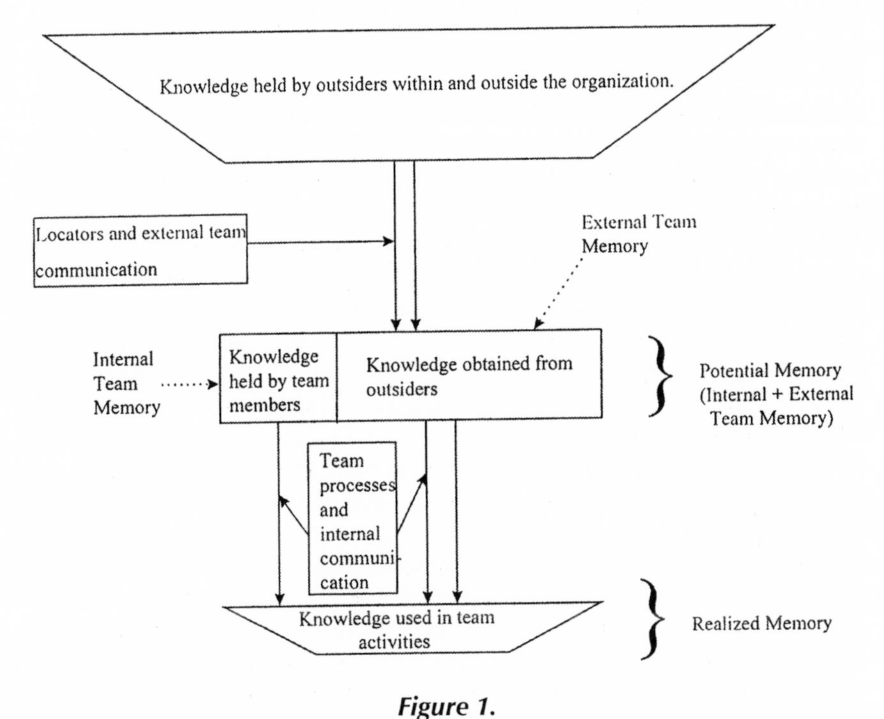

Figure 1.

realized memory of a team is the amount of potential knowledge that a team is actually able to use in its activities. These two knowledge forms and the processes that shape them are discussed below.

The Potential Memory of a Team

The potential memory of the team comprises all the knowledge held in the internal and external memories of its members. The internal memory of an individual is the knowledge which is personally known to him or her. External memory on the other hand, is knowledge not personally known to the individual, but accessible from other sources (Anand et al., 1998; Harris, 1978; Wegner, 1986; Wegner, Guiliano, & Hertel, 1985). Examples include: coworkers, associates with other companies, libraries, friends, internet resources, members, and newsletters of trade organizations.

There is an enormous amount of knowledge around the individual, and hence the question arises as to what knowledge should be considered part of a team member's external memory. This answer is provided by examining team member *directories*. Directories are idiosyncratic bundles of information that notify a given team member about the existence, location, and means for retrieving externally located information and knowledge (Wegner, 1986). Directories also include labels which are the "tags" used to name and classify chunks of information (e.g., accounts receivable is a tag that names all information relating to creditors). Thus, knowledge and information are part of an individual's external memory, if and only if the individual has appropriate directories which facilitate access to such information.

Directories are generally maintained with respect to chunks or classes of knowledge—referred to as domains—rather than with respect to specific items (Wegner, 1986). For instance, instead of maintaining directory items about specific tax laws, team members will likely maintain a single directory item pertaining to the knowledge domain of taxation. Thus, a member's directory may identify a particular associate in the accounting department as the holder of tax-related information (i.e., tax directory), rather than as holder of knowledge about a specific tax law. However, this domain-specific directory item ensures that large amounts of tax-related knowledge are included in the member's external memory.

A final component of directories are the *labels* that group members use to identify and name chunks of knowledge (Wegner, 1986). Labels are idiosyncratically held and play a critical role in facilitating information retrieval (Anand et al., 1996). For instance, Cooper and colleagues (Cooper, Folta, & Woo, 1995) point out that U.S. managers were often unable to obtain certain kinds of operational data from Japanese managers, because the two sets of managers labeled the data differently. When group members use the same labels to tag knowledge, they can more easily access information from each other.

Directories, however, are not the only factors that can determine the contents of the external memories of team members. The widespread prevalence of corporate intra-nets and networked computers has made it much easier to locate holders of expertise in organizations (Anand et al., 1998; Senna, 1997; Zorn, Marshall, & Paned, 1997). For instance, a team member requiring help on a specific domain can broadcast a public request to other organization members. Research has shown that the holders of relevant expertise generally step forward and help those who seek help (Constant, Sproull, & Keisler, 1996).

Similarly, corporate intranets are now used to store information such as the expertise held by various individuals, the projects they are currently working on, and even store data and documents related to ongoing work (Kovel, Quirk, & Gabin, 1996; Zorn et al., 1997). A team member seeking specific knowledge can use search engines to locate the holders of desired knowledge. In fact, the popular business press has described a firm that provides organizations with software that identifies specialists on a given domain of knowledge within and outside the organization (Stewart, 1997).

Thus, search processes supported by advanced electronic technologies (i.e., electronic search processes), can often link team members to external knowledge and information sources. As such, electronic search processes are viable alternatives to directories. Hence, the external memory of a team member includes knowledge and information residing outside the team, but accessible to at least one team member through a *locator,* which may either be his or her directory, *or* an electronic search process.

Apart from the presence of directories, a team member must necessarily communicate with the external knowledge holder to obtain the information. If the member is aware of the location of the knowledge, but cannot obtain it from the holder, then such knowledge cannot be considered a part of his or her external memory (Anand et al., 1996).

In addition to receiving the external knowledge, the team member must obtain such knowledge accurately. This requires the team member to be media sensitive, that is, have an ability to choose a communication medium that is appropriate to the knowledge being transferred (Daft & Lengel, 1986). Thus complex and equivocal information is transferred using rich media such as face-to-face communication, while the use of lean media (e.g., mail and memos) is appropriate for communication of unequivocal and simple information (Daft & Lengel, 1986).

Analogous to an individual's memory, a work team's external memory is composed of all the knowledge that exists in the external memories of its members, while a team's internal memory contains all the knowledge personally known to its team members. Hence, as shown in Figure 1, the amount of knowledge that is included in the team's external memory is moderated by the locators available to team members and by their communication characteristics.

The Realized Memory of a Team

The potential memory contains the maximum amount of knowledge available for use by the team, while the realized memory represents the amount of knowledge that is actually retrieved for use from the team's potential memory. Retrieval from potential memory is largely a function of the team's internal communication (i.e., communication only between team members). There is a vast body of literature that examines a variety of team member behavioral strategies and their impact on communication. For instance, supportive and/or shared leadership (Manz, 1992; Manz & Sims, 1993), participative climate (Cummings, 1978; Lawler, 1986; Lawler, 1992), use of team mental imagery and self-talk (Manz & Neck, 1995), use of appropriate behaviors and strategies during team meetings (Manz, Neck, Mancuso, & Manz, 1997; Mulvey, Veiga, & Elsass, 1996), have all been theorized to improve knowledge sharing in teams. The above topics have been discussed in detail in the literature, and we avoid their full length discussion. However, we do emphasize the fact that team members need to adopt a number of appropriate behavioral and communication strategies in order for a team to realize a large proportion of its potential knowledge.

The preceding argument makes an assumption that the team members are aware of the relevance of the knowledge they possess, and given the proper environment, will share it. This is true to a large extent, but it does not fully capture the richness of a team approach, where members may cue each other to bring forth relevant knowledge. For instance, consider the following:

> During a meeting to discuss packing norms for auto parts, Harry felt that there may be a possibility of using blister packs for certain kinds of valves and pistons. Recalling that Sally had dealt with blister pack design in her earlier job, he sought her input. During the ensuing discussion, Tom (a third team member) recalled that a couple of months ago, he had received samples of a piston in blister packs from a German supplier. Excusing himself, he went to his office, and returned with the samples and the appropriate technical specifications.

The above example illustrates that knowledge is often retrieved as a result of the interactions among team members, both formal and informal, which is an important benefit of using a team approach. However, such knowledge retrieval requires that team members possess detailed directories about each other's skills and knowledge. Familiarity with appropriate labels facilitates this process, as does the use of "rich" communication media such as face-to-face conversation (Daft, Lengel, & Trevino, 1987).

Team members today are often geographically dispersed, and deal with large amounts of complex knowledge. The realized knowledge of teams under such circumstances can be enhanced by the use of group decision support systems (Nunnamaker et al., 1996). For instance, members of cross-national teams can communicate their thoughts in their native language, while the software translates the communication for other members (Nunnamaker et al., 1996). Other group software support systems have been designed that facilitate anonymous participation, multi-criteria decision making, and brainstorming (Boland, Tenkasi, & Te'eni, 1996; Connolly, Jessup, & Valacich, 1990; Dennis, Valacich, & Nunnamaker, 1990).

In addition to ensuring that relevant knowledge from other organization members can be retrieved into the team's realized memory, it is equally important to ensure that a team retrieves knowledge held by non-organization members. Thus, individual team members facilitate the transfer of knowledge both internal to the team (i.e., between team members), as well as between external sources and the team context. Indeed, the ability of a work team to draw on knowledge from outside the organization can often be a source of organizational success (Fites, 1996; Macdonald, 1995). For instance, R&D teams in the pharmaceutical industries share knowledge with scientists in other organizations on an informal basis; moreover, team members considered that knowledge obtained through such means was crucial to the success of their teams (Leibiskind, Oliver, Zucker, & Brewer, 1996).

EMPOWERING TEAMS WITH KNOWLEDGE

Organizations that need to empower their work teams with knowledge need to perform two distinct sets of activities: (1) activities that are aimed at expanding the potential knowledge of teams, and (2) activities that are aimed at expanding the realized knowledge of teams. These activities are described below.

Organizational Initiatives That Improve a Team's Potential Knowledge

Potential team knowledge has two fundamental components: the knowledge held by the team members themselves (internal knowledge), and the knowledge held by outsiders linked to the team members through locators (external knowledge). Thus, organizational practices can build team potential memory by expanding either or both of these components, in addition to increasing the linkages between team members and knowledgeable outsiders.

Training and Team Potential Knowledge

Organizations that invest in employee training and have genuine development programs increase potential memory by increasing the personal skills and knowledge base of the employee pool; new skills and information obviously increase the knowledge available in the internal team memories. Training and development programs also increase potential memory by expanding a team's external memory. For example, if training is conducted in groups, individuals are likely to interact, share, and thus become familiar with the experiences, skills, and expertise of other program participants within the same company (Liang, Moreland, & Argote, 1996). This increases their awareness of external knowledge sources and thus external memory within their employing organization. Additionally, it also acquaints employees with each other, making it easier for a team member to seek knowledge from other organizational employees. All of this has the effect of making team boundaries more permeable and/or expanded. Different employees outside of the formal team essentially become an extension of the team itself on an as-needed basis.

An interesting and related way in which a number of industry leaders are expanding the knowledge available with work teams involves training key people external to their own firms (note that *training* generally refers to formally designed learning programs and does not include more informal on-the-job training). For instance, Toyota Motor Corporation and Chrysler Corporation have focused on training their vendors. Both organizations take the lead in ensuring that employees of key suppliers are aware of the latest engineering and management techniques. Apart from the obvious benefits with respect to the quality of services and products supplied, the increased knowledge of such outsiders increases the external knowledge of teams in these organizations. Teams inside

Toyota and Chrysler have access to larger amounts of knowledge, because they can tap the enhanced knowledge of these key outsiders. Further, as the organization has taken the lead in training the suppliers, they are more likely to respond to requests for knowledge from the various teams within the organization.

On a similar note, team members also interact with social acquaintances, and often acquire knowledge/information relevant to their tasks (Granovetter, 1982; Macdonald, 1995). Thus, knowledge held by social acquaintances of members is clearly part of the team's external memory. The greater the knowledge and expertise of individuals in the social environment of a team member, the greater the knowledge and information in the external memories of the team. It follows therefore, that teams operating in locations where the average level of education and expertise is high, also possess larger amounts of knowledge in their potential memories. In a geographic sense, communities in silicon valley and those surrounding Research Triangle Park provide examples of locales rich with high-tech and bio-tech knowledge and information. Thus computer and bio-tech firms operating in these areas possess larger amounts of knowledge in their potential memories.

The fact that a team's knowledge is often held by individuals outside the organization has another important implication. When organizations invest in improving the skills of the residents of their community, they indirectly improve the knowledge available in their potential memories. For instance, Daimler Benz has started employing more interns than it can absorb, with the hope that the skills imparted to such interns will make it easier for them to find jobs (Taylor, 1997). While discharging its role as a socially responsible organization, Daimler Benz also derives another benefit: the interns trained by the firm, would develop cordial relationships with Daimler's employees and retain a measure of good will toward the organization. At a later date, if a Daimler employee attempted to seek informal knowledge input from the intern, it would be readily provided.

Proposition 1. The knowledge in the potential memories of teams is enhanced in organizations that provide training to their employees.

Proposition 2. The knowledge in the potential memories of teams is enhanced in organizations that train their employees in groups, rather than individually.

Proposition 3. The knowledge in the potential memories of teams increases when their organizations focus on training and developing key stakeholders such as suppliers and distributors.

Proposition 4a. The knowledge in the potential memories of teams increases when their organizations are located in regions populated by skilled and knowledgeable individuals.

Proposition 4b. The knowledge in the potential memories of teams increases when their organizations initiate efforts to increase the expertise of the populace in the surrounding community.

Team Member Diversity and Team Potential Knowledge

Diversity in work teams has recently become the focus of tremendous academic and practitioner interests (as evidenced by the recent special issue of the *Academy of Management Review* [Jackson, 1996], devoted to diversity in the workplace). Team member diversity can result from variations in the age, gender, race, education, and functional background of team members. An organization promoting diversity in its teams increases team potential knowledge in two ways.

First, when team members differ from each other they tend to have been exposed to different environments. This leads them to hold differing perspectives and worldviews, which greatly increase the diversity of soft knowledge among team members (Fiske & Taylor, 1984; Waller, Huber, & Glick, 1995; Walsh, 1995). The potential knowledge of the team is thus augmented. In addition, diversity with respect to education, training, and knowledge increases the amount of knowledge available in the internal memories of teams.

Second, when team members are diverse, they tend not to socially interact with each other (Larkey, 1996). This leads them to have mutually exclusive social networks (Granovetter, 1982). Consequently, as each member brings to the team his or her unique social network, the number of outsiders linked to the team is expanded leading to a corresponding increase in the team's potential memory. Further, when the network of outsiders linked to the organization is large, the probability that the same knowledge will be accessed from multiple unrelated sources increases. This improves the reliability of the knowledge available, as members can compare knowledge items obtained from multiple sources to check for any contradictions (Anand et al., 1996).

Proposition 5. The knowledge in the potential memory of a team increases with the diversity of its members.

Organizational Socialization Practices and Team Potential Knowledge

The external knowledge of a team is expanded only when its members have appropriate directories about the knowledge held by outsiders. Directories not only inform members about the location and means of retrieval of externally stored knowledge, but they should also provide them with labels that will enable them to successfully obtain it. For instance, there exists a unique terminology or jargon within clinical medicine. The same can be said of the legal profession (i.e., "legalese"). If team members are not familiar with the appropriate labels (or jargon) they will find it extremely difficult to retrieve required knowledge.

Organizations can develop the directories of their employees by using appropriate socialization practices that are aimed at introducing and linking employees with other knowledgeable individuals (Anand et al., 1998; Van Maanen & Schein, 1979). Socialization practices may involve a formal orientation program for organizational newcomers to acquaint them with other knowledge holders, and/or a formal program which requires newcomers to spend a certain amount of time in various functional departments (Saks & Ashforth, in press). As team members can derive knowledge from individuals within and outside the organization, it is important to ensure that socialization processes acquaint organization members with both kinds of knowledge holders.

An important means through which organizations can socialize their members with outsiders is through supporting and participating in trade shows and conferences. This provides employees with exposure to information and individuals outside of the employing organization and augments both the internal and external memories of teams. The former is expanded when employees acquire personal skills and are then placed in a work team. The team's external memory is expanded when team members develop new relationships and increase their familiarity with different sources of knowledge.

Job rotation is yet another form of socialization process that extends a team's potential memory by expanding the knowledge in the rotatee's internal memory (Nonaka & Takeuchi, 1995). For example, if an employee is rotated every 18 months through different departments, then at each stop the employee's knowledge is enriched with the information and skills acquired within that particular department. Thus, any team of which this employee is a member, has the benefit of this internal memory. Additionally, the team member also develops links with employees in other departments, and these links subsequently expand a team's external knowledge.

Less formal, but often effective means of expanding potential memory are the use of internal newsletters and memos (Anand et al., 1998). These often highlight achievements of individuals and provide details of projects that may not be common knowledge to most employees. The dissemination of such information heightens the awareness of employees and thus enlarges their external memory. In a similar manner, company social gatherings (e.g., picnics, happy hours, and national meetings) provide opportunities for employees to mix with others that they may not have for during the course of a workday. This provides each with greater knowledge about the skills and knowledge of other company employees, thereby expanding external and potential knowledge. The importance of these "informal" (i.e., not directly related to specific organizational business) social events should not be underestimated. These activities allow employees to build relationships with others that are not required through the normal course of work. This, in turn, increases the potential knowledge available in their teams.

Proposition 6. The knowledge in the potential memories of teams is increased in organizations that adopt extensive socialization tactics aimed at making employees aware of each other's knowledge.

Proposition 7. The knowledge in the potential memories of teams is increased in organizations that adopt extensive socialization tactics aimed at introducing team members to knowledgeable outsiders.

History of Team Approach and Team Potential Knowledge

An important factor that leads to improved directories is the practice of using temporary cross-functional teams to perform nonrecurring and recurring organizational tasks such as developing a new product, dealing with a labor incident, and so forth (Clark, Amundson, & Cardy, 1997). By the very nature of teams, members are required to interact with one another. This forces team members to become aware of each other's expertise, which enriches their respective directories. This is especially true in the case of cross-functional teams (CFTs). When the team members of CFTs accomplish the assigned tasks and return to their normal job functions, they still retain the directories formed while working together (Anand et al., 1998). When such members subsequently participate in other teams, they bring a large amount of knowledge to each, by virtue of their highly developed directories. Thus, the use of cross-functional teams leads to a self-reinforcing loop—that is, the longer a team approach has been employed in the organization, the greater the potential knowledge available to its teams.

Proposition 8. The knowledge in the potential memories of teams is increased when their organizations have a longer history of using cross-functional teams.

Information Technology and Team Potential Knowledge

The widespread use of information technology and its application to groups has been the focus of significant research attention (Nunnamaker et al., 1996). The use of information technology (and especially group decision support systems) can improve a team's potential knowledge in two ways. First, effective group decision support systems ensure that all meeting records, past discussions, and related data are available to members at any point in time (Stein & Zwass, 1995). Further, information technology increases member perceptions that knowledge is readily available (Huber, 1990; Leidner & Elam, 1995), which in turn increases the motivation of members to locate and use such knowledge (Hertel, 1988). Thus, when work teams are supported by information technology, not only is knowledge more readily available, team members make greater efforts to locate knowledge they would otherwise not have used.

A second way in which information technology increases the potential knowledge of the team is by providing an effective substitute for the directories of members. Most organizations have installed networked computers and many have even established intranets (something akin to the internet, except that the knowledge involved is stored within the organization and available only to organization members). These intranets often maintain listings of employees, along with their areas of specialization (Zorn et al., 1997). This allows members to identify the likely holders of desired knowledge using software search engines that search the organizational computer network for documents labeled with previously defined key words (Senna, 1997; Zorn et al., 1997). Search engines can substitute synonyms and translations of key words to cast a broader net (Anand et al., 1998). Thus, search engines are not limited to the prospective labels assigned at the time of storage, but can search at a much more detailed and a broader level. Teams operating in organizations that have adopted information technology such as those described above, clearly increase the amount of knowledge in their potential memories by having access to larger amounts of externally stored knowledge.

However, merely adopting information technology and group decision support systems does not guarantee that the potential knowledge of teams in that organization is increased (Laudon & Laudon, 1994; Nunnamaker et al., 1996). It is extremely important that team members are trained to utilize information technology and the group decision software support systems. In the absence of effective training, team members may perceive the technology to be too complex and difficult which may cause them to avoid using it altogether (Cronan & Douglas, 1990; Joshi, 1991), resulting in no benefits to the potential knowledge of the team.

Proposition 9. The knowledge in a team's potential memory is increased when organizations adopt advanced information technology and group decision support systems.

Proposition 10. The relationship between team potential knowledge and information technology is moderated by the amount of information technology training provided to group members, such that at higher levels of training the relationship is positive and stronger.

Expanding the Realized Knowledge of Teams

Once a large amount of knowledge has been included in the firm's potential memory it must be transferred to the team's realized memory. A variety of factors may prevent this from happening: for instance, poor communication among team members, a lack of motivation to access knowledge from the team's external memory, a lack of awareness that their knowledge is relevant to the task at hand, and so forth. The amount of knowledge transferred from a team's potential memory to its realized memory depends on the following: (1) the organizational structure and norms, (2) team member diversity, (3) the extent of behavioral training provided to team members, and (4) the extent of communication training provided to team members.

Organizational Norms and Realized Knowledge in Teams

A very large component of a team's potential knowledge is held by other individuals linked to team members. The greater the amount of external knowledge that can be transferred to a team's realized memory, the better its performance (Ancona & Caldwell, 1992). However, a team is likely to use large amounts of its external knowledge only when its members believe that such knowledge can be obtained in a timely manner (Ancona & Caldwell, 1992). Thus, in bureaucratic organizations, smaller amounts of knowledge will be transferred to the team's realized memory because of team member perceptions that such knowledge is difficult to obtain.

The above argument clearly applies to information. However, an entirely different set of problems is encountered when we focus on soft knowledge. Soft knowledge, by definition, is not easily communicable. In order to ensure that a team can access soft knowledge in its external memory, the organizational norms of participation need to be flexible. Thus, when a team encounters a problem that requires it to seek or utilize the soft knowledge held in its external memory, its members can adopt one of the following two tactics: (1) reverse the traditional flow by communicating the problem to appropriate soft knowledge holders and seek from them a solution that the team can apply to its task, and (2) by use of the team member directories to involve knowledgeable outsiders in the team on an ad-hoc basis (Anand et al., 1998).

The above actions are facilitated when the organization has an organic structure with a minimum amount of rules and procedures. Hence, if team members require an opinion from an expert in another division, they are likely to make the effort to access such knowledge if it can be easily obtained (e.g., over an informal lunch discussion). On the other hand, if they are required to seek a formal meeting through their supervisor, they are less likely to do so.

Finally, knowledge will not be transferred to the team's realized memory if its members do not make the efforts to spontaneously retrieve such information when required. Under such conditions, the use of self-leading teams as opposed to other forms of teams (Manz, 1992), is likely to be more effective. In self-leading teams, the emphasis is on allowing team members to control their own behavior, and surrounding environment (Manz, 1986). Members of self-leading teams are responsible for assessing, selecting, and utilizing knowledge available to them in their internal and external memories. Since the shift of control to team members motivates them to greater efforts (Sims & Manz, 1996), they are more likely to make efforts to transfer knowledge from the team's realized memory.

Proposition 11. The realized memories of teams are augmented when the organization has an organic structure with little bureaucratic control.

Proposition 12. The realized memories of teams are augmented when the organization promotes self-leading teams over other forms of teams.

Team Member Diversity and Realized Knowledge in Teams

In the earlier section we had made the point that diversity increases the knowledge available in the potential memory of teams. However, when we focus on the realized memory, the situation is somewhat different. For instance, Cox (1993) has suggested that increasing diversity in a team reduces both the frequency of communication and the effectiveness of communication (his focus was on race and cultural diversity, but his arguments have subsequently been extended to apply to gender and age also). This occurs because individuals from diverse backgrounds adopt different communication styles that are often misunderstood. Further, individuals of different ethnic or racial backgrounds use different labels to classify and tag knowledge, and this complicates the process of knowledge sharing. Support for Cox's arguments is provided by Zenger and Lawrence (1989) who found that age diversity was negatively related to communication frequency in groups. Additionally, a study that focused on top management teams found that heterogeneity of work experience was negatively related to the extent of informal communication within the group (Smith, Smith, Olian, Sims, O'Bannon, & Scully, 1994).

Since the model of team memory holds that communication within the group is essential for transferring knowledge to the team's realized memory, diversity within the team leads to a reduced amount of knowledge in the realized memory of work-teams. This provides an interesting issue for managers, because diversity has two opposite effects on the knowledge availability in teams.

Proposition 13. The knowledge in the realized memories of teams is enhanced for teams consisting of homogenous members.

Behavioral Training and Realized Knowledge in Teams

There has been extensive research that has identified factors which prevent open communication in work groups. These factors include groupthink (Janis, 1983), domination by one or two individuals (Manz et al., 1997; Mulvey et al., 1996), and dysfunctional conflict (Amason, 1996). At the same time, researchers have proposed a variety of ways in which such obstacles can be overcome. These include, processes involved in "teamthink" and the use of positive thought patterns (Manz & Neck, 1995), the use of constructive conflict (Amason, 1996), and the use of appropriate conflict resolution techniques (Manz et al., 1997).

Most work team researchers agree that in order to ensure free communication between team members, a certain set of behavioral and social skills must be learned (Lawler, 1986, 1992; Sims & Manz, 1996). However, such skills are not necessarily learned over time; team members have to be trained in such skills (Mulvey et al., 1996). Especially in knowledge-intensive tasks, team members can improve performance if they have been provided the necessary skills that enable a free and unbiased exchange of ideas.

Proposition 14. The knowledge included in team realized memories increases when organizations provide team members with social and behavioral skills that enhance communication within the work group.

Communication Training and Realized Knowledge in Teams

Earlier, we pointed out that when team members interact with each other, the team's communication process can cue members to retrieve knowledge. However, such retrieval is dependent on the use of rich media such as face to face communication. Given the fact that there is an increasing tendency in organizations to have virtual teams (this is especially true of knowledge-based teams), with spatially separated members, it becomes an imperative for organizations to ensure that teams get an opportunity to meet regularly on a face-to-face basis (Anand et al., 1998).

However, in today's fast-changing environments, it is impractical to have all knowledge and information sharing through time-consuming rich media. Indeed, recent group decision support systems have greatly facilitated communication (especially of unequivocal and simple information) among spatially distributed team members (Boland et al., 1996). Thus, calling a meeting to provide team members for routine progress reports, or for any purpose where collaborative effort is not required, can be counter-productive and lead to a waste of time. In recent conversations with members of software development teams in Arizona, we were surprised at the large number of firms which called meetings sharing simple information that could have been disseminated through e-mail. Hence, efficient transfer of knowledge to a team's realized memory requires that team members judiciously choose the appropriate media when they wish to share or obtain knowledge.

Unfortunately, organization members often do not use appropriate media (Trevino, Daft, & Lengel, 1990). Various factors such as top management preference for a specific media (Markus, 1994), or social influence (Fulk, Schmitz, & Steinfeld, 1990), may result in employees being media insensitive. Thus, there is a need to train employees to be media sensitive (i.e., match the communication media to the nature of the message being transmitted). Teams in organizations that train team members to be media sensitive ensure the availability of larger amounts of knowledge in their realized memories.

Additionally, even if team members do possess directories linking them to outside knowledge holders, often they do not utilize them (Anand et al., 1998; Fites, 1996). For instance, Anand and colleagues (1998) cite the following example:

> A paper manufacturer changed the wood used to palletize (a form of packaging) its paper in an Asian country. Unfortunately the wood proved susceptible to attack by local insects, who bored first through the wooden pallets and then through the paper packaged in them. Tons of paper were scrapped. During discussions with a distributor (renowned for his packaging knowledge), managers were told: "...if only you had asked us....One of your competitors used the same wood years ago and suffered the same fate...." (Anand et al., 1998, p. 800)

The example demonstrates the need for organizations to train their members about the importance of retrieving externally located knowledge. Unfortunately, employees often are not provided with the resources and opportunities necessary to enable them to identify and access relevant knowledge. Typically, this awareness is left to experience or chance. In other words, little direction is provided as to whom to ask or when. Fortunately, a teams approach naturally facilitates knowledge awareness and exchange by expanding potential knowledge via expanded external sources, as described previously. Additionally, systems and services are emerging, both internal to organizations and on a contractual basis, that accumulate, organize, and make relevant information (both internal and external) more readily available (e.g., Stewart, 1997). In either case, employees need adequate training in order to become proficient with such resources.

Proposition 15. The knowledge included in the realized memories of teams is enhanced when the organization trains its employees to be media sensitive.

Proposition 16. The knowledge included in the realized memory of teams is enhanced when the organization trains employees and sensitizes them to the value of externally located knowledge.

IMPLICATIONS FOR RESEARCH AND PRACTICE

As we move deeper into a knowledge-based society, the need to use a teams approach in organizations is a necessity rather than an option. However, as knowledge increasingly becomes a foundation for organizational competitive advantage, the relevant question is: how can organizations maximize the knowledge available within teams? We have tried to address this issue and have argued that the organization needs to perform several supporting activities to ensure that knowledge is available within teams. While this argument is not new, we have identified support activities which had hitherto not been associated with teams. For instance, we have pointed out the relevance of socialization practices for the success of a teams approach in a knowledge-based environment, the need to examine the impact of information technology and the importance of media sensitivity in team members. We have also highlighted the importance of developing the external networks of team members as a contributor to team knowledge, as well as the importance of training both employees and key outsiders.

One of the most important implications of the preceding discussion is that a team's knowledge can be understood in terms of its memory. It is possible to identify the impact of organizational activities on team knowledge by examining their impact on the potential and realized memories of teams. Clearly, an increased understanding of team knowledge can be arrived at by integrating current research on work teams with the insights into organizational memory provided by organization theorists. For instance, researchers can gain more insight on the impact of a firm's culture, ecology, and structure on team knowledge from the work of Nelson, Winter, and Walsh (Nelson & Winter, 1982; Walsh, 1995; Walsh & Ungson, 1991). The impact of communication and information technology on team knowledge can be assessed by integrating the work teams literature with research that has taken a communication approach to group and organizational memory (Anand et al., 1998; Liang, et al., 1995; Stein & Zwass, 1995; Wegner, 1986; Wegner, Erber, & Raymond, 1991; Wegner et al., 1985). Finally, researchers trying to understand the role of soft knowledge in work teams will find fruitful leads by integrating their work with that of Karl Weick (Daft & Weick, 1984; Weick, 1995; Weick & Roberts, 1993) and that of Ikujiro Nonaka (Nonaka, 1994; Nonaka & Takeuchi, 1995).

An obvious next step in studies of knowledge in work teams is the empirical testing of the various propositions. We envisage two possible difficulties in making this step. The first problem relates to difficulties in the operationalization of knowledge. While some authors have tried to measure knowledge (Bohn, 1994; Miller & Shamsie, 1996), such operationalizations have been focused on very specific forms of knowledge. For instance, Miller's study looked at knowledge in terms of technical knowledge of Hollywood studios. Despite their limitations, these studies provide a first step for researchers attempting to measure team knowledge.

A team's potential knowledge (which consists of internal and external knowledge) could also be measured indirectly by using proxies such as tenure at job, and educational level. The external knowledge of the team may be assessed through a network analysis approach, combined with a study of the communication characteristics of team members. Considering the inherent difficulties of network analysis on large samples, this approach would be best suited for studies using small samples. Researchers wanting to work with larger samples would be best served by seeking insights from work that has examined boundary spanners in organizations and groups (Ancona & Caldwell, 1990; Tushman & Scanlon, 1981a, 1981b).

A second possible problem that we envisage in empirical studies of team knowledge relates to the levels issues. In this chapter we have made the case, that in order to ensure knowledge availability with work teams, organizations will need to carry out a large number of activities. Many of these activities (for instance, socialization tactics, training of key stakeholders), are not directed at specific work teams; rather they are global organizational activities which benefit all work teams. Thus, empirical studies will need to examine the relationship between variables at different organizational levels—an analysis that tends to get messy. Overcoming this problem will require researchers to clearly specify the level of their theory and variables and to use appropriate data analytical techniques (Glick, 1985; Klein, 1994; Ostroff, 1993).

CONCLUSIONS

Many successful organizations of the twenty-first century will implement a teams approach for performing knowledge-intensive work. The service- and information-oriented industries of the time will make this a strategic necessity for many, not merely an option. However, adopting a teams approach by itself is not going to be enough. Organizational leaders will need to take actions that ensure that their teams are provided with relevant knowledge. Since knowledge has emerged as the foundation for competitive advantage, those organizations that perform the above task better than their competitors will be the leaders of the next century.

Based on a model of team memory, we have identified a variety of activities that can be adopted to ensure knowledge availability with teams. The memory model suggests that organizations can increase knowledge availability in work teams by training key outsiders like suppliers and distributors, by adopting socialization tactics that link team members to individuals within and outside the organization and by providing behavioral and communication training to team members. Further, work teams in organizations that use appropriate information technology, adopt flexible work norms, and use a teams approach for longer periods, will have access to larger amounts of knowledge.

In summary, many of these activities are not directed at specific teams. Rather they are aimed at preparing the organization to adopt a teams approach that maximizes knowledge availability. A clear insight that emerges from the model is the increasing interdependence among team members and other individuals: specifically, the knowledge available to a team increases when the knowledge of other employees, key stakeholders, and even individuals in their community increases.

REFERENCES

Agor, W.H. (1991). How intuition can be used to enhance creativity in organizations. *Journal of Creative Behavior, 25,* 11-19.

Amason, A.C. (1996). Distinguishing the effects of functional and dysfunctional conflict on strategic decision making: Resolving a paradox for top management teams. *Academy of Management Journal, 39,* 123-148.

Anand, V., & Manz, C.C. (1997). *Knowledge empowerment in the executive suite: An organizational memory approach.* Paper presented at the Academy of Management Annual Meeting, Boston, MA.

Anand, V., Manz, C.C., & Glick, W.H. (1998). An organizational memory approach to information management in organizations. *Academy of Management Review.*

Anand, V., Skilton, P.F., & Keats, B.W. (1996). Reconceptualizing Organizational Memory. Paper presented at the Academy of Management Annual Meeting, Cincinnatti, OH.

Ancona, D.G., & Caldwell, D.F. (1990). Information technology and work groups: The case of new product teams. In J. Galegher, R.F. Kraut, & C. Egido (Eds.), *Intellectual teamwork* (pp. 302-319). Hillsdale, NJ: Lawrence Erlbaum Associates.

Ancona, D.G., & Caldwell, D.F. (1992). Bridging the boundary: External activity and performance in organizational teams. *Administrative Science Quarterly, 34,* 634-664.

Anderson, J.R. (1983). *The architecture of cognition.* Cambridge, MA: Harvard University Press.

Ashcraft, M.H. (1989). *Human memory and cognition.* Glenview, IL: Foresman and Company.

Bandura, A. (1986). *Social foundations of thought and action: A social-cognitive view.* Englewood-Cliffs, NJ: Prentice Hall.

Bazerman, M.H. (1997). *Judgment in managerial decision making* (4th ed.). New York, NY: John Wiley and Sons.

Bohn, R.E. (1994, Fall). Measuring and managing technological knowledge. *Sloan Management Review,* 61-73.

Boland, R.J., Tenkasi, R.V., & Te'eni, D. (1996). Designing information technology to support distributed cognition. In J.R. Meindl, C. Stubbard, & J.F. Porac (Eds.), *Cognition within and between organizations* (pp. 245-280). Thousand Oaks, CA: Sage.

Caldwell, B. (1996, Dec. 23). Hiring drive is on at GM. *Informationweek, 18.*

Clark, M.A., Amundson, S.D., & Cardy, R.L. (1997). Individual and organizational learning through cross-functional teams. *Proceedings of the Decision Sciences Institute, 1.*

Conner, K.R., & Prahalad, C.K. (1996). A resource based theory of the firm: Knowledge vs. opportunism. *Organization Science, 7,* 477-501.

Connolly, T., Jessup, L.M., & Valacich, J.S. (1990). Effects of annonymity and evaluative tone on idea generation in computer mediated groups. *Management Science, 36,* 689-703.

Constant, D., Sproull, L., & Keisler, S. (1996). The kindness of strangers: The usefulness of electronic weak ties for technical advice. *Organization Science, 7,* 119-135.

Cooper, A.C., Folta, T.B., & Woo, C. (1995). Entrepreneurial information search. *Journal of Business Venturing, 10,* 107-120.

Cox, T.H. (1993). *Cultural research in organizations: Theory, research and practice.* San Francisco: Berrett-Koehler.

Cronan, T.P., & Douglas, D.E. (1990). End-user training and computing effectiveness in public agencies: An empirical study. *MIS Quarterly, 6,* 14-28.

Cummings, T. (1978). Self-regulated work groups: A socio-technical synthesis. *Academy of Management Review, 3,* 625-634.

Daft, R.L., & Huber, G.P. (1987). How organizations learn: A communication perspective. *Research in the Sociology of Organizations, 5,* 1-36.

Daft, R.L., & Lengel, R.H. (1986). Organizational informational requirements, media richness and structural design. *Management Science, 32,* 554-571.

Daft, R.L., Lengel, R.H., & Trevino, L.K. (1987). Message equivocality, media selection, and manager performance: Implications for information systems. *MIS Quarterly, 11,* 355-366.

Daft, R.L., & Weick, K.E. (1984). Towards a model of organizations as interpretive systems. *Academy of Management Review, 9,* 284-295.

D'Aveni, R.A. (1994). *Hypercompetition: Managing the Dynamics of Strategic Maneuvering.* New York: Free Press.

Dennis, A.R., Valacich, J.S., & Nunnamaker, J.F. (1990). An experimental investigation of small, medium and large groups in an electronic meeting system environment. *IEEE System, Man and Cybernetics, 25,* 1049-1057.

Fiske, S.T., & Taylor, S.E. (1984). *Social cognition.* Reading, MA: Addison-Wesley.

Fites, D.V. (1996). Make your dealers your partners. *Harvard Business Review, 74*(2), 84-97.

Fulk, J., Schmitz, J., & Steinfeld, C.W. (1990). A social influence model of technology use. In J. Fulk & C.W. Steinfeld (Eds.), *Organizations and communication technology* (pp. 117-140). Newbury Park, CA: Sage.

Glick, W.H. (1985). Conceptualizing and measuring organizational climate: Pitfalls in multi-level research. *Academy of Management Review, 10,* 601-616.

Granovetter, M. (1982). The strength of weak ties: A network theory revisited. In R. Collins (Ed.), *Sociological theory* (pp. 201-233). San Francisco: Jossey-Bass.

Grant, R.M. (1996). Prospering in dynamically-competitive environments: Organizational capability as knowledge integration. *Organization Science, 7,* 375-387.

Harris, J.E. (1978). External memory aids. In M.M. Gruneberg, P.E. Morris, & R.N. Sykes (Eds.), *Practical aspects of memory* (pp. 31-38). London: Academic Press.

Hertel, P.T. (1988). Monitoring external memory. In M.M. Gruneberg, P.E. Morris, & R.N. Sykes (Eds.), *Practical aspects of memory: Current research and issues.* New York: John Wiley and Sons.

Huber, G.P. (1990). A theory of the effects of advanced information technologies on organizational design, intelligence and decision making. *Academy of Management Review, 15,* 47-91.

Huber, G.P., & Glick, W.H. (1993). Sources and forms of organizational change. In G.P. Huber & W.H. Glick (Eds.), *Organizational change and redesign: Ideas and insights for improving performance* (pp. 3-15). New York: Oxford University Press.

Jackson, S.E. (Ed.). (1996). Special topic forum on diversity within and between organizations [Special issue]. *Academy of Management Review, 21.*

Janis, I.L. (1983). *Groupthink.* Boston: Houghton Mifflin.

Joshi, K. (1991). A model of users' perspective on change: The case of information systems technology implementation. *MIS Quarterly, 15,* 187-199.

Klein, J.A. (1994). Maintaining expertise in multi-skilled teams. In M.M. Beyerlein & D.A. Johnson (Eds.), *Advances in interdisciplinary studies of work teams* (Vol. 1, pp. 145-166). Greenwich, CT: JAI.

Kovel, J., Quirk, K., & Gabin, J. (1996). *The lotus notes idea book.* Reading, MA: Addison-Wesley Publishing Co.

Larkey, L.K. (1996). Toward a theory of communicative interactions in culturally diverse work groups. *Academy of Management Review, 21,* 463-491.

Laudon, K.C., & Laudon, J.P. (1994). *Management information systems: Organizations and technology* (3rd ed.). New York: Macmillan College Publishing.

Laudon, K.C., & Starbuck, W.H. (1996). Organizational information and knowledge. In M.W. (Ed.), *International encyclopedia of business and management* (Vol. 4, pp. 3923-3933). London: Thompson Press.

Lawler, E.E. (1986). *High involvement management.* San Francisco: Jossey Bass.

Lawler, E.E. (1992). *The ultimate advantage: Creating the high involvement organization.* San Francisco: Jossey Bass.

Leibiskind, J.R., Oliver, A.L., Zucker, L., & Brewer, M. (1996). Social networks, learning and flexibility: Sourcing scientific knowledge in new bio-technology firms. *Organization Science, 7,* 428-443.

Leidner, D.E., & Elam, J.J. (1995). The impact of executive information systems on information on organizational design. *Organization Science, 6,* 645-664.

Liang, D.W., Moreland, R., & Argote, L. (1996). Group versus individual traning and group performance: The mediating role of transactive memory. *Personality and Social Psychology Bulletin, 21,* 384-393.

Macdonald, S. (1995). Learning to change: An information perspective on learning in the organization. *Organization Science, 6,* 557-568.

Machlis, S. (1997, Feb. 24). Still looking for a few good CIOs. *Computerworld, 31,* 28.

Manz, C.C. (1986). Self-leadership: Towards an expanded theory of self influence processes in organizations. *Academy of Management Review, 11,* 585-600.

Manz, C.C. (1992). Self-leading work teams: Moving beyond self-management myths. *Human Relations, 45,* 1119-1139.

Manz, C.C., & Neck, C.P. (1995). Teamthink: Beyond the groupthink syndrome in self-managing teams. *Journal of Managerial Psychology, 10,* 7-15.

Manz, C.C., Neck, C.P., Mancuso, J., & Manz, K.P. (1997). *For team members only: Making your work place productive and hassle free.* New York: Amacom.

Manz, C.C., & Sims, H.P. (1980). Self management as a substitute for leadership: A social learning perspective. *Academy of Management Review, 5,* 361-377.

Manz, C.C., & Sims, H.P. (1993). *Business without bosses: How self-managing teams are building high performing organizations.* New York: Wiley.

Manz, C.C., & Stewart, G.L. (1997). Attaining flexible stability by integrating total quality management and socio-technical systems theory. *Organization Science, 8,* 59-70.

Markus, M.L. (1994). Electronic mail as the medium of managerial choice. *Organization Science, 5,* 502-527.

Miller, D., & Shamsie, J. (1996). The resource based view of the firm in two environments: The Hollywood film studios from 1936 to 1965. *Academy of Management Journal, 39,* 519-543.

Mulvey, P.W., Veiga, J.F., & Elsass, P.M. (1996). When teammates raise a white flag. *Academy of Management Executive, 10,* 40-48.

Nelson, R.R., & Winter, S.G. (1982). *An evolutionary theory of economic change.* Cambridge, MA: The Belknap Press.

Nonaka, I. (1994). A dynamic theory of knowledge creation. *Organization Science, 5,* 14-36.

Nonaka, I., & Takeuchi, H. (1995). *The knowledge creating company.* New York: Oxford University Press.

Nunnamaker, J.F., Briggs, R.O., Mittleman, D.D., Vogel, D.R., & Balthazar, P.A. (1996). Lessons from a dozen years of group support systems research: A discussion of lab and field findings. *Journal of Management Information Systems, 13,* 163-207.

Ostroff, C. (1993). Comparing correlations based on individual-level and aggregated data. *Journal of Applied Psychology, 78,* 569-582.

Polanyi, D.M. (1966). The tacit dimension. London: Routledge and Kegan Paul.

Quinn, J.B. (1993). *Intelligent Enterprise.* New York, NY: Free Press.

Sackman, S.A. (1992). Culture and sub-cultures: An analysis of organizational knowledge. *Administrative Science Quarterly, 37,* 140-161.

Saks, A.M., & Ashforth, B.E. (in press). Organizational Socialization: Making sense of the past and the present as a prologue for the future. *Journal of Vocational Behavior.*

Senna, J. (1997, June 9). Phantom indexes your intranet. *Infoworld, 19,* 74.

Sims, H.P., & Manz, C.C. (1996). *Company of heroes: Unleashing the power of self-leadership.* New York: John Wiley & Sons.

Smith, K.G., Smith, K.A., Olian, J.D., Sims, H.P., O'Bannon, D.P., & Scully, J.A. (1994). Top management team demography and process: The role of social integration and communication. *Administrative Science Quarterly, 39,* 412-438.

Spender, J.C. (1996). Making knowledge the basis of a dynamic theory of the firm. *Strategic Management Journal, 17,* 45.

Stein, E.W., & Zwass, V. (1995). Actualizing organizational memory with information systems. *Information Systems Research, 6*(2), 85-117.

Stewart, T.A. (1997, September 29). Does anyone around here know...? *Fortune, 136,* 279-280.

Susman, J.I. (1976). *Autonomy at work: A socio-technical analysis of participative management.* New York: Praeger.

Taylor, A. (1997). Neutron Jurgen ignites a revolution at Daimler Benz. *Fortune, 136*(9), 144-154.

Trevino, L.K., Daft, R.L., & Lengel, R.H. (1990). Understanding managers' media choices: A symbolic interactionist perspective. In J. Fulk & C. Steinfeld (Eds.), *Organizations and communication technology* (pp. 71-94). Newbury Park, CA: Sage.

Tushman, M.L., & Scanlon, T.J. (1981a). Boundary spanning individuals: Their role in information transfer and their antecedents. *Academy of Management Journal, 24,* 83-96.

Tushman, M.L., & Scanlon, T.J. (1981b). Characteristics and external orientations of boundary spanning individuals. *Academy of Management Journal, 24,* 289-305.

Van Maanen, J., & Schein, E.H. (1979). Toward a theory of organizational socialization. In B.M. Staw & L.L. Cummings (Eds.), *Research in Organizational Behavior* (Vol. 1, pp. 209-264).

Waller, M.J., Huber, G.P., & Glick, W.H. (1995). Functional background as a determinant of executives selective perception. *Academy of Management Journal, 38,* 873-896.

Walsh, J.P. (1995). Managerial and organizational cognition: A trip down memory lane. *Organization Science, 6,* 280-321.

Walsh, J.P., & Ungson, G.R. (1991). Organizational memory. *Academy of Management Review, 16,* 57-91.

Wegner, D.M. (1986). Transactive memory: A contemporary analysis of the group mind. In B. Mullen & G.R. Goethals (Eds.), *Theories of group behavior.* New York: Springer-Verlag.

Wegner, D.M., Erber, R., & Raymond, P. (1991). Transactive memory in close relationships. *Journal of Personality and Social Behavior, 61,* 923-929.

Wegner, D.M., Guiliano, T., & Hertel, P. (1985). Cognitive interedependence in close relationships. In W.J. Ickes (Ed.), *Compatible and incompatible relationships.* New York: Springer-Verlag.

Weick, K.E. (1995). *Sensemaking in organizations.* Thousand Oaks, CA: Sage.

Weick, K.E., & Roberts, K.F. (1993). Collective mind in organizations: Heedful interrelating on flight decks. *Administrative Science Quarterly, 38,* 357-381.

Zenger, T.R., & Lawrence, B.S. (1989). Organizational demography: The differential effects of age and tenure distributions on technical communication. *Academy of Management Journal, 32,* 353-376.

Zorn, P., Marshall, L., & Paned, M. (1997, May/June). Surfing corporate intranets: Search tools that control the undertow. *Online, 21,* 30-51.

COGNITIVE ENGINEERING IN TEAM PERFORMANCE MEASUREMENT

Renée J. Stout, Nancy J. Cooke, Eduardo Salas, Dana M. Milanovich, and Janis A. Cannon-Bowers

ABSTRACT

The purpose of this chapter is to specify what team performance measures should assess and how cognitive engineering techniques can uncover this information. This chapter will briefly review what is known about teams, teamwork, and team performance. Following this is a discussion of team performance measurement in terms of its purposes and requirements, what the cognitive engineering approach is all about, as well as a description of specific methods that have been derived through this approach for understanding the underlying cognitive components of performance. This chapter concludes with a description of some examples of how the cognitive engineering approach can be used to develop team performance measures.

Advances in Interdisciplinary Studies of Work Teams, Volume 6, pages 27-53.

ISBN: 0-7623-0655-6

INTRODUCTION

Today, more than ever, the workplace is characterized by dynamic and rapidly changing task conditions. Moreover, technological changes in modern work environments have drastically increased the cognitive complexity of work. As a result, the cognitive demands placed upon operators necessitates the use of teams of individuals working in concert to effectively complete their jobs (Cannon-Bowers, Salas, & Converse, 1993; Stout, Cannon-Bowers, & Salas, 1996/1997). These teams must work together in an efficient and coordinated manner, because incidents in which teams implement poor team processes have the potential to cause catastrophic results.

Indeed, there have been many documented examples of the dire consequences of breakdowns in team coordination in a variety of operational environments. For example, within a military aviation setting, teamwork failure contributed to the February 1992 crash of a transportation helicopter. The crew members were focused on a failed engine and lost awareness of other elements of the situation. Specifically, while focusing on one engine's failure, the crew failed to execute proper engine restart procedures, causing the second engine to fail, as well. Also, the crew's failure to instruct the passengers exacerbated this already catastrophic situation. Two lives were lost, as was the aircraft. Had the crew implemented more effective teamwork processes, this tragedy could potentially have been avoided.

Given the criticality of teams and teamwork to the accomplishment of real world tasks, there is a need to derive efficacious methods, tools, and strategies that enable teams to gain and maintain the requisite competencies for effective integrated task performance (Cannon-Bowers & Salas, 1997). Cannon-Bowers, Tannenbaum, Salas, and Volpe (1995) specifically pointed to the need to translate teamwork into necessary knowledge, skills, and attitudes (KSAs) as a requirement for the design of training programs. A cornerstone to designing and implementing training to enhance these teamwork KSAs is to develop measurement approaches that reliably and validly assess their underlying components. Howell and Cooke (1989) pointed out the importance of understanding cognitive factors underlying performance in the context of training research. Other researchers within the team performance arena have also indicated the need for such a cognitive focus (Hall & Regian, 1996; Salas, Bowers, Cannon-Bowers, 1995).

Unfortunately, even though several researchers have indicated that measurement tools must address cognition, one of the predominant issues associated with the state-of-the-art in team performance measurement concerns the use of overly simplistic measures of behavior and the concomitant lack of attention paid to the cognitive factors underlying behavior and performance. In this chapter, we argue that a cognitive engineering approach to deriving measures of team performance can overcome many of the limitations of existing team performance measures.

The purpose of this chapter is to specify *what* team performance measures should assess and how cognitive engineering techniques can uncover this information. We will start by briefly reviewing what is known about teams, teamwork, and team performance. Following this, we discuss team performance measurement in terms of its purposes and requirements. Next we discuss the cognitive engineering approach and describe specific methods that have been derived through this approach for understanding the underlying cognitive components of performance. We then provide some examples of how the cognitive engineering approach can be used to develop team performance measures. We conclude by briefly discussing some of the practical implications of this approach.

TEAMS

Although teams have been important to the accomplishment of a variety of operational tasks for many years, it has only been in recent years that a better understanding of what constitutes a team has emerged. Part of the problem has been definitional in nature. That is, the term *team* has often been confused with the term *group* in the literature (Brannick, Roach, & Salas, 1991). As such, the available literature on teams has resulted in ambiguous, confusing, and conflicting findings leading to few integrated interpretations of the resultant body of studies (Stout, Salas, & Fowlkes, 1997). Therefore, we feel it necessary to provide a definition of teams here to establish the boundaries of the propositions that are put forth in this chapter.

The definition of a team provided by Salas, Dickinson, Converse, and Tannenbaum (1992) seems to capture the essence of a myriad of other definitions given by different authors (e.g., Boguslaw & Porter, 1962; Briggs & Naylor, 1964; Dyer, 1984; Hall & Rizzo, 1975; Morgan, Glickman, Woodard, Blaiwes, & Salas, 1986; Nieva, Fleishman, & Reick, 1978). Specifically, the Salas and colleagues' (1992) definition emphasizes the need for interaction and interdependency, leading to requirements for sharing of information and the engagement of cooperative action. Therefore, we will use the definition of Salas and colleagues here, that defines a team as "a distinguishable set of two or more people who interact, dynamically, interdependently, and adaptively toward a common and valued goal/objective/mission, who have each been assigned specific roles and functions to perform, and who have a limited life-span of membership" (p. 4).

We should point out that a set of two or more individuals are often required to engage in the cooperative interaction suggested by Salas and colleagues (1992), yet do not employ this teamwork. These individuals are still a team—they are simply an ineffective team. "Thus, the crux of teamwork is for the team members to use a collection of processes, strategies, and actions that allow them to effectively and efficiently perform"(Stout, Salas, & Fowlkes, 1997, p. 170). We now turn to a discussion of teamwork and team performance.

Teamwork and Team Performance

Historically, the majority of studies conducted in the team area focused on characteristics of teams, such as cohesion, group homogeneity/heterogeneity, group size, cognitive ability of members, and group structure rather than on the interactional requirements placed upon teams that must coordinate their activities (Hackman, 1976). As a result, the literature has provided little guidance on what teams actually must *do* to achieve effective performance.

The lack of attention that has been paid to the investigation of interaction requirements is surprising given that many authors have noted the importance of interactional processes in teams (Bass, 1980; Brannick, Roach, & Salas, 1993; Hackman, 1976; Hackman & Morris, 1975; McIntyre & Salas, 1995; Prince & Salas, 1993; Stout, Salas, & Carson, 1994). In essence, these researchers have indicated the necessity for team members to be proficient at their individual task accomplishment, but that this is not sufficient to ensure effective team performance. Thus, they have noted that the effectiveness of team performance is also dependent upon the quality of interactions among team members.

Several researchers have identified potential interactional processes that are necessary for effective team performance (McIntyre & Salas, 1995; Nieva et al., 1978; Prince & Salas, 1993; Smith-Jentsch, Salas, & Baker, 1996). Many of the proposed processes are specific behaviors that teams are assumed to be required to engage in for effective performance. As such, there exists the potential to observe whether or not a team exhibits these behaviors. Once this is established, it is then possible to empirically determine the relationship between specific behaviors and performance. Also, a review of this literature reveals that the behaviors appear to represent several different domains or dimensions of process behaviors, thus providing a fairly thorough delineation of team processes.

Only in the past decade have empirical investigations begun to provide a better understanding of the processes necessary for attaining effective performance. Indeed, results of several recent studies suggest that a variety of processes appear to be important for effective team performance (e.g., Brannick et al., 1993; McIntyre & Salas, 1995; Stout et al., 1994). Measures of teamwork, therefore, must consider multiple behavioral dimensions (Baker & Salas, 1997). However, as noted by Cannon-Bowers and colleagues (1993), the majority of the empirical investigations on team processes have essentially ignored the more elusive components of teamwork, such as adaptability and coordination of action, and have instead focused on more readily observable components, such as communication. As a result, existing team performance measurement schemes tend to be behaviorally focused and simplistic. Other researchers have also indicated the need to investigate the cognitive components of team performance in terms of underlying knowledge (Cannon-Bowers et al., 1995; Cooke, Stout, & Salas, 1997; Hall & Regian, 1996; Howell & Cooke, 1989), as well as to derive team performance measures sensitive to cognitive factors of team performance (Cooke et al., 1997;

Kraiger, Ford, & Salas, 1993; Stout, Salas, & Fowlkes, 1997; Stout, Salas, & Kraiger, 1997).

Fortunately, a recent comprehensive review of the literature outlined in this chapter and of several other research efforts in the area of team performance, was conducted by Cannon-Bowers and colleagues (1995). Through an integration of this literature, Cannon-Bowers and colleagues derived the underlying competencies of teamwork, which contain both a knowledge and a skill component, as discussed above, in addition to an attitudinal component. These authors described team competencies to be the KSAs that comprise effective teamwork and team performance. Additionally, Cannon-Bowers and colleagues specified a set of knowledge, skill, and attitude competencies necessary for teams to operate successfully in a variety of different environments. This approach expands upon previous related work on individual training (e.g., Goldstein, 1986). Given the focus on team competencies, measures must therefore assess each of these factors.

Knowledge competencies are essentially the knowledge bases that team members need to possess to execute their team tasks. These include an understanding of team member roles and responsibilities (e.g., Baker, Salas, Cannon-Bowers, & Spector, 1992), cue-strategy associations (e.g., Cannon-Bowers et al., 1995), and shared mental models (e.g., Cannon-Bowers, Salas, & Converse, 1993). Skill competencies are what enable team members to actually carry out required functions. Cannon-Bowers and colleagues (1995) collapsed existing suggested team processes, as described above, and, through a card sorting procedure, developed a synthesized taxonomy of skills. This was not done as an attempt to delineate new processes, but instead as an attempt to summarize and integrate previous work. The resulting skills were adaptability, situational awareness, performance monitoring and feedback, leadership and team management, interpersonal relations, coordination, communication, and decision making, each with accompanying subskills. Finally, attitude competencies are beliefs that team members have about performing team tasks and include attitudes toward teamwork, collective orientation (e.g., Driskell & Salas, 1992), collective efficacy (e.g., Shea & Guzzo, 1987), and cohesion.

Given the importance of requisite team competencies to team performance, delineating the specific KSAs underlying the performance situation at hand becomes paramount in order to develop effective and useful team performance measurements. We argue here that cognitive engineering techniques can provide valuable insight for deriving the knowledge component of team competencies. That is, this process would specify *what* should be assessed in a team performance situation.

In this section we have argued that team performance measures must identify KSAs required in the task situation and whether or not they are being met. We now turn to a direct discussion of team performance measures, including consideration of their various purposes and specification of some requirements of team performance measures.

Team Performance Measurement

What are They Used For?

Team performance measurement information (as well as any performance measurement information) can serve a variety of purposes. These include, but are not limited to: (1) selection, (2) program evaluation, (3) trainee performance appraisal or evaluation, (4) training, and (5) research purposes, in order to better understand the nature of constructs studied. Furthermore, there are subcategories of purposes that can be established for each of the five general purposes listed above. For example, in the training arena, performance measurement information can be used to provide feedback to the trainee, to diagnose performance for subsequent instruction to the trainee, or to examine the causes of performance as a starting point for restructuring an entire training program. We suggest that cognitive engineering techniques can be useful for uncovering information relevant to each of the purposes and sub-purposes of team performance measurement. We argue, however, that cognitive engineering approaches are particularly useful for developing team performance measures used for training purposes, because these require a richer understanding of underlying processes and the causes of these processes in team settings—which is precisely what cognitive engineering techniques are best suited toward discovering. Therefore, we next consider some specific requirements that team performance measures in training settings must address.

What Must Team Performance Measures in Training Do?

According to Cannon-Bowers and Salas (1997), team performance measures used for training purposes must meet several criteria. They must: (1) consider multiple levels of measurement; (2) address processes as well as outcomes; (3) be able to describe, evaluate, and diagnose performance; and (4) provide a basis for remediation. Because the first and second criteria drive the third and fourth, we will elaborate upon them below.

Performance measurement can be targeted to individuals or teams (Cannon-Bowers & Salas, 1997; Prince, Brannick, Prince, & Salas, 1997; Stout, Prince, Salas, & Brannick 1995). Examples of behaviors at the individual unit of analysis are: (1) the leader clearly reassigned duties as appropriate (e.g., "you fly and I'll troubleshoot"); (2) the First Officer volunteered to do tasks without being asked (e.g., "would you like me to set up for the landing now?"); and (3) the First Officer correctly identified a warning light (e.g., "it's the left boost pump"). Examples of team level performance include: (1) the team sequenced or timed their activities well; (2) the team shared information when arriving at decisions; and (3) as an example of poor performance, neither team member assumed the responsibility for making decisions, resulting in hesitant action on the part of the team. Depending on the purpose of measurement, it may be more appropriate to focus at a particular level of analysis. For some purposes, such as training, however, measurements should be taken at both the individual and team unit of analysis, because both are important to enhancing training.

Performance measures can also target outcomes or processes (Cannon-Bowers & Salas, 1997; Smith-Jentsch, Johnston, & Payne, 1998). Outcomes are the end results or choices made in task accomplishment. An example of an outcome is whether or not a flight team avoided icing conditions. Outcomes are usually fairly objective in nature in that a given outcome did or did not occur, but tend to provide little or no diagnostic information that can be used to help teams improve their performance. Processes are the steps taken by the team and the behaviors engaged in by the team that allowed for the accomplishment of a given outcome. For any given outcome, the processes needed for accomplishment may range from being precise and well defined (e.g., the sequence required to arm a missile) to being more ambiguous, loosely defined, and affected by contextual factors (e.g., which systems a team brought back on-line following an electrical fire, the order in which they brought them on, how they discussed this with one another, and who they informed of their activities). In addition, processes can be the steps, sequence, and activities required to complete the task, such as standard operating procedures (SOPs) or normal operating procedures, or they can involve more of the teamwork types of behaviors involved in completing the job. Examples of the latter are whether the team avoided talking over each other and generally backed each other up appropriately.

While process interactions tend to be more difficult to rate than outcomes, for some purposes, such as training, it is critical that assessment approaches focus on capturing relevant process behaviors. This is because systematic identification of needed processes is the crux in improving performance in the task domain. For training, in complex, dynamic task situations which impose ambiguous cues, emphasis should be placed on teaching teams how to make decisions correctly rather than on making the correct decision (Cannon-Bowers & Salas, 1997). The latter can occur due to chance operations of the team and should not necessarily be reinforced in training.

Given this discussion, we contend that team performance measures, especially when used for training, must assess requisite KSAs held at the individual, as well as team level. Moreover, these measures must be able to provide feedback not only on individual and team outcomes, but on specific processes used by individuals and by teams to successfully complete their tasks. Thus, cognitive engineering techniques should be used to identify measures to assess each of four components: (1) individual outcomes, (2) team outcomes, (3) individual processes, and (4) team processes (see Cannon-Bowers & Salas, 1997, for greater detail). For purposes of remediation, which is the crux of team performance measurement in training, uncovering specific individual and team processes used in effective and ineffective task execution is essential. Further, the causes associated with using effective and/or ineffective processes is critical to accurate diagnosis. Cognitive engineering techniques, with their emphasis on real tasks in real settings, can uniquely contribute to the development of team performance measures, especially for training purposes, by identifying these processes and their corollary causes. However, before providing an illustration of how the cognitive engineering approach can aid in the development of team performance measures, however, we will describe cognitive engineering.

THE COGNITIVE ENGINEERING PERSPECTIVE

We begin this section by stating our belief that taking a cognitive engineering approach to defining problems represents a particular *perspective* that is analogous to the perspective taken in the field of naturalistic decision making (NDM) (Klein, Orasanu, Calderwood, & Zsambok, 1993). That is, many articles have been written on the NDM position that research must focus on what effective decision makers actually do (Zsambok & Klein, 1997). A focus on actual decision making behavior is critical, because the available empirical evidence strongly suggests that classical decision making models have no direct relevance to real world decisions (Klein, 1989). Thus, the NDM perspective takes a strong position on studying real tasks in real contexts to determine the underlying strategies that lead to effective decision making and team performance. Likewise, simply attending to the title of the book written by Hutchins (1995), that champions the cognitive engineering approach—*Cognition in the Wild*—makes one quickly aware that researchers taking a cognitive engineering approach also advocate studying cognition outside of the lab. As a result of this perspective, the field of cognitive engineering has generated a number of new methods that are better suited for capturing the richness and intricacy of cognition as it is applied to real and complex problems (e.g., Christoffsen, Hunter, & Vincente, 1994; Klinger & Gomes, 1993; McNeese, 1995).

We should note that, on the one hand, the cognitive engineering perspective is more inclusive than the NDM perspective, in that it covers cognitive phenomena other than decision making. On the other hand, NDM is more inclusive in that it represents a movement or philosophy in a general sense for studying decision making, and it utilizes the tools that have been generated by cognitive engineers for uncovering cognition in decision making settings. Given an understanding of the perspective of cognitive engineering, it is important to describe the purpose of this approach.

What is the Purpose of Cognitive Engineering?

Cognitive engineering brings theories, principles, and data from cognitive psychology to applied, real-world problems (Cooke et al., 1997). Cognitive engineering attempts to understand the perceptual, attentional, memory, and decision making structures and processes underlying complex tasks for the design of systems, such as intelligent tutoring systems and human-computer interfaces and the development of training programs (Hutchins, 1995; McNeese, 1995; Woods & Roth, 1988).

A highly related field, knowledge engineering, also involves the process of building intelligent software, such as expert systems and intelligent tutors, but it focuses on the construction of a knowledge base that is at the heart of these systems (Cooke, 1994; Cooke et al., 1997). Therefore, knowledge engineering concentrates on the knowledge component of cognition, as opposed to considering perceptual processes, for example. Others have argued that the etiology of the methods used in the two fields differs and that knowledge engineering has been primarily technology driven, whereas cognitive engineering has been user driven (McNeese, 1995). Thus, there are subtle distinctions between the two approaches, however, integral to both cognitive and knowledge engineering are the use of knowledge elicitation techniques (Cooke et al., 1997). We turn now to explaining the process of knowledge elicitation.

Knowledge Elicitation

Knowledge elicitation is "the process of collecting from a human source of knowledge, information that is thought to be relevant to that knowledge" (Cooke, 1994, p. 802). Elicited knowledge has often been criticized due to its limitations. Cooke (1994) described several of these limitations. For example, data gathered through verbal reports has been criticized for being incomplete and inaccurate (Nisbett & Wilson, 1977). Experts are often unable to verbalize their knowledge due to knowledge compilation and chunking (Anderson, 1982; Kraiger et al., 1993). In other instances, they may distort their knowledge, apply only textbook examples, or may attempt to satisfy the needs of the elicitor (Cooke, 1994). Others, have indicated instances in which elicited knowledge is quite valid (Ericsson & Simon, 1984). Therefore, accurately and effectively implementing knowledge elicitation techniques to derive underlying knowledge related to human performance is crucial.

A recent comprehensive effort by Cooke (1994) reviewed and integrated the literature on knowledge elicitation from a variety of sources including psychology, business management, education, counseling, cognitive science, linguistics, philosophy, and anthropology, in an attempt to provide guidance regarding when to apply particular techniques as a function of the purpose of the application. However, she did not specifically address the application of these techniques to team settings and to deriving team performance measures. Rather, she categorized knowledge elicitation techniques and discussed the specific strengths and weaknesses of each approach. We will next draw heavily upon her review to describe what we see as the basic categories of knowledge elicitation techniques, and we will briefly describe their associated advantages and disadvantages. In addition we will briefly discuss some issues to consider when using these techniques scientifically in team environments in terms of what they can contribute, as a sample illustration. Since our intention is to provide a snapshot of how the various techniques can be used, we will not address all of the knowledge elicitation techniques that can be applied to team environments. The reader is referred to Cooke (1994) for a more comprehensive review of the techniques available.

Categories of Knowledge Elicitation Techniques

Knowledge elicitation techniques can be broadly categorized as observations, interviews, process tracing methods, and conceptual methods (Cooke, 1994; Cooke et al., 1997). Each category contains many variations and generally differs in terms of procedure, type of knowledge elicited, and advantages and disadvantages.

Observation is not typically given enough credit as a technique for knowledge elicitation. On the contrary, it has the potential to provide vast amounts of information in the context of performance, while interfering minimally with the task. It is an ideal technique to use in initial analyses for it enables the observer to acquire a "feel" for the task, including the general procedures and decisions involved, constraints, and information requirements. Observations are typically recorded in written, audio, or video form.

Variations within the observation category of techniques revolve around issues of how to observe (i.e., passively or actively) and what to observe (i.e., predetermined features or specific aspects of the task and environment). The advantage of minimal intrusion is offset by difficulties in interpreting the data. It is difficult to draw firm conclusions from observational data. Therefore, these methods should be used to generate hypotheses or ideas or to verify information elicited in some other way.

Fortunately in response to developments in cognitive engineering, there has been a push for tools to aid in observational analysis. The goal is to capture the complexity of the situation, while at the same time understand and draw some conclusions about what was observed. For instance, there are tools available that interface video recorders with personal computers so that annotations, counts, and classifications of behaviors can be entered as the video is viewed (Harrison, 1991). Then, if desired, video clips corresponding to a specific type of event can be automatically located and played back. Also, various analyses of frequencies and transitions are common. Timelines (Owen, 1993) and MacSHAPA (Sanderson, McNeese, & Zaff, 1994) are two tools with these capabilities.

In team performance situations, it is important to understand what cues team members attend to and share in order to form both accurate individual understandings of the situation and a shared understanding of the situation among team members (Stout et al., 1996/1997). Observation would be a useful method for deriving some preliminary ideas about the informational cues in the environment affecting each team member's understanding of the situation. The nonintrusive nature of this method enables the observed task to proceed as naturally as possible, which is critical for identifying subtle cues that are used. Other methods for identifying cues are not as forgiving in this regard, and thus, observational methods may provide information on how the task is distorted by the use of more intrusive techniques.

Interviews are the most frequently employed of all knowledge elicitation techniques. This is probably because, like observations, they are excellent for gaining a general understanding of the situation and generating and verifying hypotheses. In addition, interviews, especially unstructured ones, can be administered with very little preparation, although it is useful for the interviewer to be practiced at facilitating conversation. Interviews are typically retrospective, and therefore subject to memory retrieval failures on the part of the expert. We also note that the use of unstructured interview techniques has suffered similar criticisms as unstructured selection interviews. Finally, like observations, the resulting data are difficult to interpret. It is a good idea to record interviews in written, audio, or video form.

Unstructured interviews are free-form in that neither content nor sequencing is prespecified. Structured interviews follow a predetermined format and can vary from highly rigid to only loosely constrained. Although structured interviews require more preparation time than unstructured interviews, they generally suffer from less severe criticisms. Structured interviews also have advantages of being more systematic, and therefore complete, and more comfortable for both participants. Most importantly, structured interviews can be sure to address specific questions, if they exist.

An added aid in conducting structured interviews is to provide the participant with paper so that he or she can make lists or even draw diagrams representing elements of the situation and their interrelations. Also, during this process, it may be useful to employ a concept mapping procedure, such as COGENT (Cognitive Engineering Network Technologies) (Sanderson et al., 1994), in which the operator expresses his or her knowledge in terms of a concept map in which concepts are represented as nodes, and relations between concepts are represented as links between nodes (Novak, 1990, 1995; Rentsch, Heffner, & Duffy, 1994).

To gain an understanding of team performance requirements, after some initial free responses, it would probably be beneficial to increase the structure in the interview. This could be done, for example, by asking the operator to report what he or she knows about specific aspects of the situation (e.g., other team members' activities). Further, information about the team task goals can be measured by starting an interview with an outcome or goal (e.g., abort landing) and asking the operator to work backwards from that outcome by listing the information that led to such a decision, followed by the information or cues that preceded that information, and so on (this is a particular form of structured interview called goal decomposition). The point here, as with the use of this technique for individual tasks, is to elicit information that may have been overlooked, while at the same time avoiding leading or biasing questions. Also, even though an individual is interviewed, a focus on team processes, interactional requirements, and team level cues ensures that team level information is elicited.

In addition, knowledge about team members can be elicited by using two other structured interview techniques. One is to have the operator draw a diagram with nodes representing individual team members and to connect those nodes with directed links indicating information flow between team members. Another technique that could be used here is a structured interview that centers on role reversal. That is, a team member assumes the position and duties of another team member during a task simulation. The interview can then focus on views of the task and system from other team members' positions. This is helpful when it is not that important that the operator have a complete understanding of the system, so this type of knowledge may be overlooked if not viewed from other team members' perspectives.

Some tailoring of techniques to meet the specifics of the task may also be required. Because many tasks of interest are well-practiced ones, it may be the case that the operator is wrong about some report of current or long-term knowledge. If this is suspected, it is wise to verify information elicited in the structured interview through either analysis of video records or the administration of queries that capture the elicited information.

Process tracing techniques are methods of collecting data specific to task performance, often concurrently with task performance. These data are later analyzed to make inferences about the thoughts or cognitive processes associated with that task performance. One of the most popular forms of process tracing relies on verbal reports provided while "thinking aloud." These reports can be retrospective, but it is typically believed that as long as there is no interference with the task itself, concurrent reports provide more valid information regarding the current contents of working memory. Specifically, concurrent reports are not subject to the memory loss problems associated with retrospective reports and are thus believed to provide a more accurate representation of one's "current" task knowledge.

In general, verbal reports have been heavily criticized on the basis of their inaccuracy and incompleteness (e.g., Nisbett & Wilson, 1977). The best advice is to use this type of information sparingly—in situations in which communication is not a large part of the task and in situations in which concurrent verbalizing will not disrupt the primary task. Also, it is often very difficult to elicit verbal reports. There has been some success in using dyads in which two individuals think aloud to each other while performing the task. Another means of facilitating verbal reports is to elicit the concurrent report as the operator views a tape of prior task performance. This may be a solution in cases in which interference appears to be an obstacle.

Nonverbal data are also collected to trace cognitive processes. These include keystrokes, actions, facial expressions, and gestures. Eye movements in particular have been used to identify cues used by experts (Carmody, Kundel, & Toto, 1984). Although there is not a one-to-one mapping between eye movements and thoughts, they can be used in conjunction with verbal reports to help verify them. For example, if an operator claims to be attending to a particular cue, but he or she never scans the instrument where that cue is located, then the accuracy of the verbal report would be called into question. Also, despite the fact that eye movement instrumentation is costly, the data it generates are useful for getting at order and weight information in sequential tasks, as well as for clarifying and verifying information (Lauwereyns & d'Ydewalle, 1996).

Protocol analysis is the term used to describe various methods for summarizing and interpreting process tracing data. It typically involves transcribing the data, developing a coding scheme that captures the critical content of the data, applying the coding scheme to each identified unit in the protocol, and exploring frequencies, patterns, and sequential dependencies in the results. A relatively new area of inquiry that focuses on exploratory sequential data analysis (ESDA) (Sanderson & Fisher, 1994) has produced some analytical tools and methods in recent years to facilitate this process. Several protocol editing and analysis tools have been developed (e.g., MacSHAPA; Sanderson et al., 1994). In addition, PRONET (Cooke, Neville, & Rowe, 1996) is a procedure for generating node-link graph structures from the co-occurrence of items in a sequence. In the case of process tracing, those items would be units of verbal reports or nonverbal behavior.

Process tracing methods can be used in at least two ways to elicit knowledge underlying team performance. First, individual concurrent verbal reports taken while interacting with a simulation (or thereafter while viewing a video of performance), should provide some evidence of the cues that are identified by operators. If these cues are visual in nature, then they might be verified using eye movements. Otherwise, verification can be accomplished by varying the values of several hypothesized cues and regressing action or decision outcomes on these values. High or statistically significant regression weights associated with cues support the hypothesis that the cue is used. Cues in the environment are probably the most difficult information to elicit, because they are often used automatically, without conscious awareness by skilled operators. Process tracing methods may, however, provide a better indication of cues used, beyond that of structured interviews, because they are concurrent and less direct. Combining the observation method with process tracing will allow a broad range of cues to be elicited which can then be narrowed by the eye movement and regression verification methods.

The second way that process tracing can be used is to analyze communication or information sharing. In this case, any time information is passed or communication occurs, an event is recorded as part of a written protocol. Later, this protocol can be analyzed using techniques, such as PRONET (Cooke et al., 1996). Analyses of this type would result in a graph of team members as nodes with links between those who communicate frequently. Note that this is different from the method of eliciting information about team members in which an operator's knowledge about team communication is elicited. In this case, the actual behavior itself is recorded and analyzed. Think aloud protocols concurrent with task performance, on the other hand, may be problematic in that they can interfere with communications that are a part of performing the task. Adaptations of the method would have to be made in team settings.

Conceptual methods are a group of techniques that produce representations of domain concepts and their relations. Methods included in this set are cluster analysis, multidimensional scaling, Pathfinder, and concept mapping. Conceptual techniques tend to be less direct than observations, interviews, and process tracing techniques and require less introspection and verbalization than these other techniques. In general, the methods take pairwise estimates of relatedness for a set of concepts and generate a spatial or graphical representation of those concepts and their relations. The overall goal of these methods is to reduce the set of distance estimates in a meaningful way. Resulting representations can then be compared qualitatively and, in many cases, quantitatively across groups and individuals.

As previously noted, these techniques tend to be indirect in that they require judgments about conceptual relatedness, as opposed to introspections or explicit verbal reports. One advantage of these methods, particularly for measuring team performance, is that they can handle results from multiple individuals, including summarizing data from several individuals and comparing data across individuals. There are methods to quantitatively compare resulting representations that are very much like correlational measures. In fact, they have been used most often to identify group or individual differences in cognitive structures. Thus, these methods can be used, for instance, to compare knowledge structures for pairs of individuals to determine the extent of overlap.

For example, once relatedness estimates have been collected, then they can be submitted to one of several algorithms, such as Pathfinder network scaling (Schvaneveldt, 1990). In the case of Pathfinder, resulting networks can be compared in terms of the proportion of shared links. Similarly, information graphs generated from sequential data on communication and analyzed using the PRONET methodology (Cooke et al., 1996), can also be compared or summarized using these methods. For instance, each team's information sharing graph could be compared to some ideal pattern, or to the patterns of two different teams, or the same team performing two tasks that differ in complexity or workload. In any case, data can be aggregated over multiple experts to generate a composite structural representation, therefore, conceptual methods tend to handle multiple experts better than observations, interviews, or process tracing techniques. However, some have argued that the resulting representations have little relationship to actual task performance (e.g., Geiwitz, Kornell, & McCloskey, 1990). In short, the conceptual methods are constrained to knowledge about concepts and their relations, but provide excellent ways to compare knowledge structures across individuals or teams.

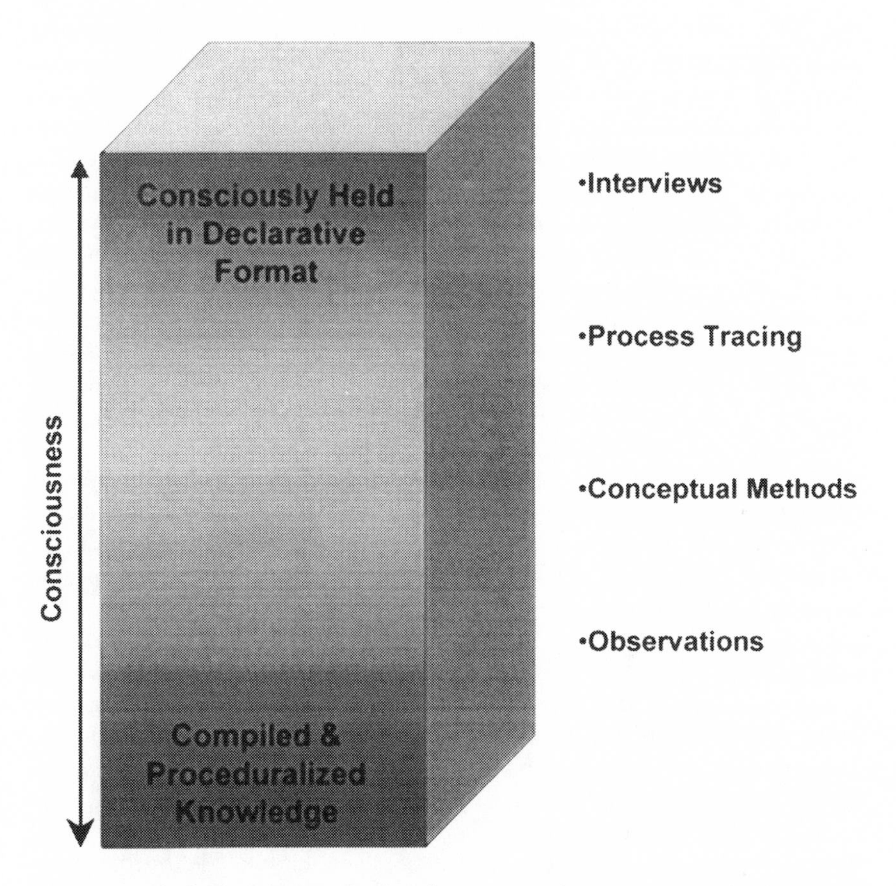

Figure 1. Cognitive Engineering Technique of Choice as a Factor of Degrees of Verbalizability of Knowledge

In summary, the four major categories of techniques as described by Cooke (1994) and Cooke and colleagues (1997) each have associated advantages and disadvantages. Some are less obtrusive than others, while some are more amenable for integrating information from various sources. Each of these techniques can be applied to the development of team performance measures. Figure 1 suggests that the choice of method used to uncover the information that should be assessed with a team performance measure should be driven by the degree to which the knowledge associated with the performance is able to be verbalized.

Extending the measurement of team performance to include content-rich applications such as training, requires that researchers add to their repertoire of assessment measures, measures from cognitive and knowledge engineering that are better suited for revealing the complex nature of the KSAs underlying effective team performance. Cooke and colleagues (1997) argued that existing measures that attempt to tap the cognitive underpinnings of performance are impoverished because of their focus on assessment instead of elicitation. In other words, they attempt to determine the degree to which some knowledge is present or absent rather than determining *what* knowledge is present and whether or not the knowledge is adequate. In addition, these authors note the importance of deriving assessment measures through knowledge elicitation so the content validity of the assessment measures will not be questionable, and the measures likely to fail. Thus, using a cognitive engineering approach can help to overcome many of the limitations of current measures by virtue of their: (1) emphasis on the cognitive underpinnings of complex tasks, (2) focus on elicitation rather than just assessment, (3) goal of studying cognition in complex environments, (4) use of qualitative, as well as quantitative data, and (5) focus on initial exploratory data analysis followed by confirmatory analysis.

The product of the elicitation process can be used to derive assessment methods that are focused on determining the degree or quality of team performance. Applications such as training evaluation and system design may require this kind of information. For example, if it is determined that teams must engage in specific processes to enable effective performance in the task situation at hand, one could determine whether or not the team actually engaged in these processes, and therefore performed effectively. The assessment methods can and should also be used to confirm that the information elicited and used to derive team performance measures is valid. This is important because despite the best efforts of the experimenter to elicit complete and unbiased information, there are no guarantees. The information provided may be ambiguous and errors of interpretation can be made by the experimenter. Thus, before applications or assessments are based on this information, the validity of the elicited information must be confirmed. In general, the confirmation procedure should collect data using the elicitation-based assessment method and then compare the results to changes in individual or team performance. If the information elicited accurately reflects team performance, then variations in the assessment measure should correspond to variation in other measures of team performance.

This discussion has provided a general review of the benefits associated with using cognitive engineering techniques to derive information that must be captured in team performance measures. We now discuss how these techniques can be used to derive measures of team performance.

USING COGNITIVE ENGINEERING TECHNIQUES TO DERIVE TEAM PERFORMANCE MEASURES

In this section, we suggest potential ways that cognitive engineering techniques can be used to enhance team performance measurement systems. However, note that our suggestions should be refined and tested in a number of environments and for various purposes (e.g., for training or performance management).

In a sense, each of the four types of techniques hold the potential to serve as measures. For example, Kraiger and Wenzel (1997) provided several illustrations of how methods such as protocol analysis and interviews could be used as team performance measures. On the other hand, information gained from the use of a particular technique serves to derive measures that fall more generally under another category of technique. For example, the concepts used in the measurement of conceptual approaches are delineated through the use of observations, interviews, and/or process tracing methods. Also, measures of accuracy, sequencing and timeliness, that might be assessed with the use of observational scales and checklists should be derived through the use of a combination of the four categories of methods. Therefore, it is highly misleading to suggest that one particular technique is best suited to derive a particular type of measure, because it may inaccurately imply that it is the only technique needed to develop the measure. Providing an in-depth discussion of all of the possible permutations of developing measures for use in particular circumstances through the combination of each of these methods is beyond the scope of this chapter. Instead, we provide a few examples of measures that can be derived which would then be considered to be generally categorized under one of the four techniques.

The methods for each of the four types of techniques described here, can potentially be used to assess completeness and accuracy of team member KSAs, as well as the extent to which team members share a common or complementary understanding of the task situation. These are characteristics which presumably lead to more effective performance (Cannon-Bowers et al., 1995). This is most directly obvious for conceptual techniques, especially as we have described them here.

For example, the Pathfinder technique mentioned earlier has two indexes particularly relevant in team settings: coherence and similarity. Coherence is an indication of internal consistency (Stout et al., 1996/1997) and thus provides some assessment of how well organized the knowledge structures of individual team members are. Similarity can be used to determine how similar an individual's knowledge structure is to a referent structure (i.e., a domain expert), and thus this gives an estimation of how accurate the knowledge structure is. Furthermore, it can be used to determine whether team members hold common knowledge structures (i.e., shared mental models). There has been very recent evidence that teams with members who hold knowledge structures more similar to an expert's and/or more similar to one another are more effective (Minionis, 1994; Stout, Salas, & Kraiger, 1997). Thus, each of these indices can be used as team performance measures for training evaluation purposes. That is, if training is effective, one's knowledge structure should become more similar to the referent (i.e., expert) structure. One can also speculate that qualitatively determining specific aspects of knowledge structures that are deficient through this type of measure can be used to provide feedback to team members, as well as to drive subsequent training.

Table 1. Summary of Knowledge Elicitation Techniques, their Purpose, and Implications for Team Performance Measurement

Technique	*Purpose*	*Implications for Team Performance Measurement*	*Source*
OBSERVATIONS	• to provide vast amount of information in the context of performance with minimal intrusion		
-TARGETs	• to obtain information on behaviors exhibited by team members during task performance	• to score team performance • to assess accuracy of process behaviors and performance outcomes • to verify the utilization of information assessed with other knowledge elicitation techniques	• Fowlkes, Lane, Salas, & Franz (1994)
INTERVIEWS	• to generate and verify ideas of hypotheses • to gain a general understanding of a situation • to gain an understanding of performance requirements	• to compare team members' responses • to interview question on performance requirements • to determine the extent of team member's expectations for others in various situations	• Kraiger & Wenzel (1997)
PROCESS TRACING	• to collect information specific to task performance • to make inferences about cognitive processes related to task performance		

-Protocol Analysis	• to acquire information on cues individuals attend to perform a task, their current and future perceptions of a situation, information they would seek and provide to fellow team members, and steps necessary to accomplish a task	• to compare to an expert referent to determine the quality of a team member's responses with respect to understanding overall situation and understanding what information sources to utilize	
-TARGETs	• to obtain information on behaviors exhibited by team members during task performance	• to provide an indication of the importance of knowledge in addition to behaviors necessary for task performance	
CONCEPTUAL METHODS	• to produce representations of domain concepts and their relations • to compare knowledge structures across individuals or teams		
-Pathfinder	• to obtain estimates of relatedness for a set of concepts and generate a graphical representation of those concepts and their relations	• to assess the organization and accuracy of a team member's knowledge structure • to determine the degree of overlap in knowledge structures among team members	• Stout, Salas, & Kraiger (1997)
		• to compare team members' knowledge structures to experts to determine team effectiveness	• Miniones (1994)

With protocol analysis, an individual team member may be asked to perform a task during which he or she would describe (1) the steps that he or she is taking, (2) the cues that he or she is attending to, (3) his or her current and future perceptions of the situation, and/or (4) information that he or she would be seeking and providing to fellow team members (while the actual seeking and passing of information would be best suited to capture through observation). Comparing responses to an expert referent would enable one to determine how closely the two sets of responses match, as an index of quality of responses. For example, if there are 10 steps necessary to complete a task, responses can be scored for degree of correctness. Similarly, if cue patterns A, B, and C are most critical in a particular part of the mission and these trigger responses X, Y, and Z, the degree to which these are articulated and acted upon can be assessed. This information can be used to evaluate training, determine trainee progress, and structure feedback to the trainee, as well as drive future instruction.

Consider, for example, that in a given scenario, ambiguous cues are presented as to what the intentions are of a particular target. Assume also that through cognitive engineering approaches, one is able to determine that this pattern signifies that the team member should seek further information from a particular piece of equipment, from a fellow crew member, and visually from the environment. Suppose through the verbal protocol elicitaion process if during the performance of the scenario, a trainee describes that he or she was assessing information from a piece of equipment other than the one that he or she should have been attending to, and he or she also fails to mention the information sought from the fellow team member and environment. This may suggest that performance in the real setting could suffer. This may be one potential indication that training that led to these incorrect responses was deficient or that the trainee was in need of remediation. Furthermore, feedback could be provided to the trainee regarding the information sources from which he or she should have sought information in this particular situation. These would be specific cognitive processes that would be difficult to identify without the use of knowledge elicitation techniques. Finally, future trainees might provide additional examples similar to the one that created difficulty for the trainee.

Caution should be made, however, when there are various paths to reaching a solution, all of which are reasonably correct. In these cases, the expert referent should establish "better" responses when possible, and these should be used in judging the quality of information provided by the team member. Fowlkes, Lane, Salas, Franz, and Oser (1994) identified a method of behavioral observation that could be adapted for use with process tracing methods and in situations in which actual observation of team members performing the task are precluded. In these cases, a checklist of "better" responses could be established and used to score protocols. This would provide an indication of knowledge versus behaviors, both of which are important measures of team performance. Also, as with conceptual methods, degree of common responses and/or complementary responses between team members can be established and linked to what is needed in the situation to determine whether team member share appropriate knowledge of the situation.

The use of interviews could follow closely the aforementioned process tracing methods used to derive team performance measures. That is, the same type of information could be elicited as described above, but rather than having the team member talk aloud, the individual would respond to a series of queries. Thus, he or she could be provided with a description of a scenario or task situation and asked questions such as: (1) "why would you perform this task?"; (2) "what should other team members be doing as you perform your duties?"; (3) "how would you know if the steps that you would take to complete the task are successful?"; (4) "what are the three most important cues that you would be attending to in this portion of the scenario"; and (5) "what would be an indication that a fellow team member is not interpreting the situation as you are?" Results could be scored in a manner similar to what was described for using process tracing methods. Also, the steps or probes could be provided to another team member, and he or she would be asked to describe the reasons for the steps, cues to determine success or failure of the steps, and so on. According to Kraiger and Wenzel (1997), "team members who could provide detailed responses to these probes would be more likely to hold accurate expectations for others' behavior in different situations" (p. 72).

With observations, observational checklists, such as targeted acceptable responses to generated events and tasks (TARGETS) (Fowlkes et al., 1994; Dwyer, Fowlkes, Oser, Salas, & Lane, 1997) could be derived and used to score performance. These are used to assess both process and outcomes in terms of accuracy of responses and can serve training, as well as evaluation purposes. Also, as previously noted, scales which measure timeliness and appropriate sequencing of activities could be used to score both individual and team performance. In general, observations can be made to verify the utilization of information assessed with each of the other methods.

Table 1 lists the four basic knowledge elicitation techniques discussed in this section along with their primary purposes. In addition, regarding the specific techniques of a particular category that were discussed in this section (e.g., protocol analysis and Pathfinder), primary purposes of the methods are listed, as are the implications of these techniques for team performance measurement.

Given the above discussion, it is critical to reiterate, that in complex and dynamic team performance settings, a multi-measurement approach should be taken by combining the use of a variety of measures derived from the cognitive engineering approach. Moreover, team performance measures developed from incorporating the use of these techniques must: (1) consider multiple levels of measurement (individual/team); (2) address processes as well as outcomes; (3) be able to describe, evaluate, and diagnose performance, including delineating underlying causes of performance and moment by moment changes; (4) provide a basis for remediation, through feedback, knowledge of results, and driving future instruction; and (5) be theoretically based, including the consideration of models of performance that incorporate cognitive, behavioral, and affective factors that are generalizable across situations. By incorporating the methods of the cognitive engineering approach, we believe that we have the potential to meet each of these criteria. Certainly, the diagnosis of underlying processes and their causes is of critical importance, since this will allow us to guide remediation.

PRACTICAL IMPLICATIONS

As with almost any approach, there is good news and bad news about applying cognitive engineering techniques to the development of team performance measures. The good news is that these techniques provide very rich and comprehensive data about the knowledge that underlies team performance. The bad news is, at least at this point, cognitive engineering techniques are labor intensive. Some of the techniques that have been described here are easier to implement (e.g., interviews) than others (e.g., conceptual models, conceptual models and interviews) and we realize that practicality issues will influence the technique one chooses to incorporate. For instance, it may not be reasonable to ask a maintenance worker to provide a verbal report of what he or she is doing while on the job, since this might distract the worker and lead to unsafe conditions. However, a practical alternative may be to generate a paper-based scenario that could be used in conjunction with an interview protocol. Using this alternative, the maintenance worker could be asked questions such as, "at this point in your task, what would you be attending to?" and "why would that be important?" which may reveal valuable information.

So, how does one go about selecting the most appropriate cognitive engineering technique? While there are a variety of factors to consider, we will briefly discuss three issues to bear in mind when making your decision. First, it is important that one considers what the information gathered will be used for *before* actually implementing a technique, because this may influence the technique chosen. For example, cognitive task analyses are too often performed without consideration of what type of information this technique will generate and whether or not this is the type of information that is needed. Second, one needs to consider the amount of time and resources available. The knowledge elicitation techniques reviewed here require that time be spent either: (1) "up-front" to carefully apply the technique to the domain of interest, or (2) "post-hoc" to make inferences about the data collected. Thus, the practitioner must make a trade-off based upon his or her purpose, the time constraints that he or she faces, and other related factors. We would recommend implementing techniques such as observation and/or interviews first and then moving on to more sophisticated techniques such as process tracing and concept mapping to uncover richer information. However, we would also stress that care should be taken to structure one's observations and/or interviews so as to improve the ease of interpretation as much as possible. A final issue is the task or job that one wants to target. That is, how dynamic and/or complex is the task? Consider for a moment, a task that is not dynamic, but rather more proceduralized in nature. While it may take less effort (e.g., time, energy, funds) to apply the cognitive engineering approach to proceduralized tasks, the gains may be only marginal, since the cognitive requirements for performing these types of tasks is low. Conversely, tasks that are highly complex are, for the most part, more cognitively demanding, more difficult to train, and it is more difficult to measure performance (especially the cognitive components). Thus, applying the cognitive engineering approach to complex tasks will require a great deal more effort, but the expected payoff will be much greater. For instance, this approach should reveal the underlying cognitive components of performance which could then be used to guide training and measurement efforts.

SUMMARY

This chapter has addressed how cognitive engineering approaches can add value in eliciting knowledge related to team performance, which is critical to consider when designing and developing team performance measures. Without considering the knowledge underlying performance, measures will continue to be simplistic and immature. We have made a case for how each of the different types of knowledge elicitation techniques can uncover important knowledge related to team performance. Finally, we have proposed sample measures that can be derived using each of the cognitive engineering approaches, but argued that a combination of methods is needed to derive most team performance measures. Not only do we hope that this stimulates research in this area in which actual team performance measures are derived and tested through the use of cognitive engineering approaches, but we must insist upon empirical tests of what has been described here before we can feel confident in what we have proposed. The potential merit to team performance measurement that we have identified in cognitive engineering, however, should invoke such interest.

NOTE

The views expressed here are those of the authors and do not reflect the official position of the organization with which the authors are affiliated.

REFERENCES

Anderson, J.R. (1982). Acquisition of cognitive skill. *Psychological Review, 89,* 369-406.

Baker, C.V., Salas, E., Cannon-Bowers, J.A., & Spector, P. (1992, April). *The effects of interpositional uncertainty and workload on team coordination skills and task performance.* Paper presented at the annual meeting of the Society for Industrial and Organizational Psychology, Montreal, Canada.

Baker, D.P., & Salas, E. (1997). Principles for measuring teamwork: A summary and look toward the future. In M.T. Brannick, E. Salas, & C. Prince (Eds.), *Team performance assessment and measurement* (pp. 331-355). Mahwah, NJ: LEA.

Bass, B.M. (1980). Individual capability, team performance, and team productivity. In E.A. Fleishman & M.D. Dunnette (Eds.), *Human performance and productivity* (pp. 179-232). Hillsdale, NJ: LEA.

Boguslaw, R., & Porter, E.H. (1962). Team functions and training. In R.M. Gagne (Ed.), *Psychological principles in systems development* (pp. 337-416). New York: Holt, Rinehart, & Winston.

Brannick, M.T., Roach, R.M., & Salas, E. (1991). *Understanding team performance: A multimethod study.* Paper presented at the meeting of the Society of Industrial and Organizational Psychology, St. Louis, MO.

Brannick, M.T., Roach, R.M., & Salas, E. (1993). Understanding team performance: A multimethod study. *Human Performance, 6,* 287-308.

Briggs, G.E., & Naylor, J.C. (1964). *Experiments on team training in a CIC-type task environment* (NAVTRADEVCEN Tech. Rep. No. 1327-1). Port Washington, NY: United States Naval Training Device Center.

Cannon-Bowers, J.A., & Salas, E. (1997). A framework for developing team performance measures in training. In M.T. Brannick, E. Salas, & C. Prince (Eds.), *Team performance assessment and measurement* (pp. 45-62). Mahwah, NJ: LEA.

Cannon-Bowers, J.A., Salas, E., & Converse, S.A. (1993). Shared mental models in expert team decision making. In N. Castellan, Jr. (Ed.), *Current issues in individual and group decision making* (pp. 221-246). Hillsdale, NJ: LEA.

Cannon-Bowers, J.A., Tannenbaum, S.I., Salas, E., & Volpe, C.E. (1995). Defining team competencies and establishing team training requirements. In R. Guzzo & E. Salas (Eds.), *Team effectiveness and decision making in organizations* (pp. 333-380). San Francisco: Jossey Bass.

Carmody, D.P., Kundel, H.L., & Toto, L.C. (1984). Comparison scans while reading chest images: Taught but not practiced. *Investigative Radiology, 19,* 462-466.

Christoffsen, K., Hunter, C.N., & Vincente, K.J. (1994). Cognitive "dipsticks": Knowledge elicitation techniques for cognitive engineering research. *Cognitive Engineering Laboratory Technical Report No. 94-01.* Toronto, Canada: University of Toronto, Department of Industrial Engineering.

Cooke, N.J. (1994). Varieties of knowledge elicitation techniques. *International Journal of Human-Computer Studies, 41,* 801-849.

Cooke, N.J., Neville, K.J., & Rowe, A.L. (1996). Procedural network representations of sequential data. *Human-Computer Interaction, 11,* 29-68. Santa Monica, CA: Human Factors Society.

Cooke, N.J., Stout, R.J., & Salas, E. (1997). Broadening the measurement of situation awareness through cognitive engineering methods. *Proceedings of the Human Factors and Ergonomics Society 41st Annual Meeting* (pp. 215-219).

Driskell, J.E., & Salas, E. (1992). Collective behavior and team performance. *Human Factors, 34,* 277-288.

Dwyer, D.J., Fowlkes, J.E., Oser, R.L., Salas, E., & Lane, N.E. (1997). Team performance measurement in distributed environments: The TARGETs methodology. In M.T. Brannick, E. Salas, & C. Prince (Eds.), *Team performance assessment and measurement* (pp. 137-154). Mahwah, NJ: LEA.

Dyer, J.L. (1984). Team research and team training: A state-of-the-art review. In F.A. Muckler (Ed.), *Human factors review: 1984* (pp. 285-323). Santa Monica, CA: Human Factors Society.

Ericsson, K.A., & Simon, H.A. (1984). *Protocol analysis: Verbal reports as data.* Cambridge, MA: MIT Press.

Fowlkes, J.E., Lane, N.E., Salas, E., Franz, T., & Oser, R. (1994). Improving the measurement of team performance: The TARGETs methodology. *Military Psychology, 6,* 47-61.

Geiwitz, J., Kornell, J., & McCloskey, B. (1990). *An expert system for the selection of knowledge acquisition techniques* (Tech. Rep. No. 785-2). Santa Barbara, CA: Anacapa Sciences.

Goldstein, I.L. (1986). *Training in organizations: Needs assessment, development, and evaluation.* Pacific Grove, CA: Brooks/Cole.

Hackman, J.R. (1976). Group influences on individuals. In M. Dunnette (Ed.), *Handbook of industrial and organizational psychology* (pp. 1455-1525). Chicago: Rand McNally.

Hackman, J.R., & Morris G.G. (1975). Group task, group interaction processes, and group performance effectiveness: A review and proposed integration. In L. Berkowitz (Ed.), *Advances in experimental social psychology* (pp.45-99). New York: Academic Press.

Hall, E., & Regian, W. (1996). *Cognitive engineering for team tasks.* Unpublished manuscript.

Hall, E.R., & Rizzo, W.A. (1975, March). *An assessment of U. S. Navy tactical team training* (TAEG Report No. 18). Orlando, FL: Training Analysis and Evaluation Group.

Harrison, B.L. (1991). Video annotation and multimedia interfaces: From theory to practice. *Proceedings of the Human Factors and Ergonomics Society 35th Annual Meeting* (pp. 319-323). Santa Monica, CA: Human Factors Society.

Howell, W.C., & Cooke, N.J. (1989). Training the human information processor: A look at cognitive models. In I. Goldstein (Ed.), *Training and development in work organizations: Frontier series of industrial and organizational psychology* (pp. 121-182). New York: Jossey Bass.

Hutchins, E. (1995). *Cognition in the Wild.* Cambridge, MA: The MIT Press.

Klein, G.A. (1989). Recognition-primed decisions. In W. Rouse (Ed.), *Advances in Man-Machine Systems Research, 5,* 47-92.

Klein, G.A., Orasanu, J., Calderwood, R., & Zsambok, C.E. (Eds.). (1993). *Decision making in action: Models and methods.* Norwood, NJ: Ablex.

Klinger, D.W. & Gomes, M.E. (1993). A cognitive systems engineering application for interface design. *Proceedings of the Human Factors and Ergonomics Society 37th Annual Meeting* (pp. 16-20). Santa Monica, CA: Human Factors Society.

Kraiger, K., Ford, J.K., & Salas, E. (1993). Application of cognitive, skill-based, and affective theories of learning outcomes to new methods of training evaluation. *Journal of Applied Psychology, 78,* 311-328.

Kraiger, K., & Wenzel, L.H. (1997). Conceptual development and empirical evaluation of measures of shared mental models as indicators of team effectiveness. In M.T. Brannick, E. Salas, & C. Prince (Eds.), *Team performance assessment and measurement* (pp. 63-84). Mahwah, NJ: LEA.

Lauwereyns, J., & d'Ydewalle, G. (1996). Knowledge acquisition in poetry criticism: The expert's eye movements as an information tool. *International Journal of Human Computer Studies, 45,* 1-18.

McIntyre, R.M., & Salas, E. (1995). Measuring and managing for team performance: Emerging principles from complex environments. In R. Guzzo & E. Salas (Eds.), *Team effectiveness and decision making in organizations* (pp. 149-203). San Francisco: Jossey-Bass.

McNeese, M.D. (1995). Cognitive engineering: A different approach to human-machine systems. *CSERIAC Gateway, 6,* 1-4.

Minionis, D.P. (1994). *Enhancing team performance in adverse conditions: The role of shared team mental models and team training on an interdependent task.* Unpublished doctoral dissertation, George Mason University, Fairfax, VA.

Morgan, B.B., Jr., Glickman, A.S., Woodard, E.A., Blaiwes, A.S., & Salas, E. (1986). *Measurement of team behaviors in a Navy environment.* (NTSC Tech. Rep. No. 86-014). Orlando, FL: Naval Training Systems Center.

Nieva, VF., Fleishman, E.A., & Reick, A. (1978). *Team dimensions: Their identity, their measurement, and their relationships* (Final Tech. Rep., Contract No. DAH19-78-C-0001). Washington, DC: Advanced Research Resources Organizations.

Nisbett, R.E., & Wilson, T.D. (1977). Telling more than we can know: Verbal reports on mental processes. *Psychological Review, 84,* 231-259.

Novak, J.D. (1990). Concept mapping: A useful tool for science education. *Journal of Research in Science Teaching, 27,* 937-949.

Novak, J.D. (1995). Concept mapping: A strategy for organizing knowledge. In S.M. Glynn & R. Duit (Eds.), *Learning science in the schools: Research reforming practice* (pp. 229-245). Mahwah, NJ: Lawrence Erlbaum.

Owen, R. (1993). *Timelines: A system for analysing time based data.* Documentation, CSRI, University of Toronto.

Prince, A., Brannick, M.T., Prince, C., & Salas, E. (1997). The measurement of team process behaviors in the cockpit: Lessons learned. In M.T. Brannick, E. Salas, & C. Prince (Eds.), *Team performance assessment and measurement* (pp. 289-310). Mahwah, NJ: LEA.

Prince, C., & Salas, E. (1993). Training and research for teamwork in the military aircrew. In E.L. Wiener, B.G. Kanki, & R.L. Helmreich (Eds.), *Cockpit resource management* (pp. 337-366). Orlando, FL: Academic Press.

Rentsch, J.R., Heffner, T.S., & Duffy, L.T. (1994). What you know is what you get from experience: Team experience related to teamwork schemas. *Group and Organizaiton Management, 19*(4), 450-474.

Salas, E., Bowers, C.A., & Cannon-Bowers, J.A. (1995). Military team research: Ten years of progress. *Military Psychology, 7,* 55-75.

Salas, E., Dickinson, T.L., Converse, S.A., & Tannenbaum, S.I. (1992). Toward an understanding of team performance and training, In R.W. Swezey & E. Salas (Eds.), *Teams: Their training and performance* (pp. 3-29). Norwood, NJ: LEA.

Sanderson, P.M., & Fisher, C. (1994). Exploratory sequential data analysis: Foundations. *Human-Computer Interaction, 9,* 251-317.

Sanderson, P.M., McNeese, M.D., & Zaff, B.S. (1994). Handling complex real-world data with two cognitive engineering tools: COGENT and MacSHAPA. *Behavior Research Methods, Instruments, & Computers, 26,* 117-124.

Schvaneveldt, R.W. (1990). *Pathfinder associative networks: Studies in knowledge organization.* Norwood, NJ: Ablex.

Shea, G.P., & Guzzo, R.A. (1987). Group effectiveness: What really matters? *Sloan Management Review, 3,* 25-31.

Smith-Jentsch, K.A., Johnston, J.H., & Payne, S. (1998). Measuring team-related expertise in complex environments. In J.A. Cannon-Bowers & E. Salas (Eds.), *Decision making under stress: Implications for individual and team training.* Washington, DC: American Psychological Association.

Smith-Jentsch, K.A., Salas, E., & Baker, D.P. (1996). Training team performance-related assertiveness. *Personnel Psychology, 49,* 909-936.

Stout, R.J., Cannon-Bowers, J.A., & Salas, E. (1996/1997). The role of shared mental models in developing team situation awareness: Implications for training. *Training Research Journal, 2,* 85-116.

Stout, R.J., Prince, C., Salas, E., & Brannick, M.T. (1995, April). *Beyond reliability: Using crew resource management (CRM) measurements for training.* Paper presented at the 10th International Symposium on Aviation Psychology, Columbus, OH.

Stout, R.J., Salas, E., & Carson, R. (1994). Individual task proficiency and team process: What's important for team functioning. *Military Psychology, 6,* 177-192.

Stout, R.J., Salas, E., & Fowlkes, J. (1997). Enhancing teamwork in complex environments through team training. *Group Dynamics: Theory, Research, & Practice, 1,* 169-182.

Stout, R.J., Salas, E., & Kraiger, K. (1997). The role of trainee knowledge structures in aviation team environments. *The International Journal of Aviation Psychology, 7*(3), 235-250.

Woods, D.D., & Roth, E.M. (1988). Cognitive systems engineering. In M. Helander (Ed.), *Handbook of human-computer interaction* (pp. 3-43). Amsterdam: Elsevier Science Publishing Company.

Zsambok, C.E., & Klein, G. (Eds.).(1997). *Naturalistic decision making*. Mahwah, NJ: LEA.

HELPING TEAM MEMBERS HELP THEMSELVES

PROPOSITIONS FOR FACILITATING GUIDED TEAM SELF-CORRECTION

Kimberly A. Smith-Jentsch, Elizabeth Blickensderfer, Eduardo Salas, and Janis A. Cannon-Bowers

ABSTRACT

The current trend toward downsizing, both in industry and the military, has necessitated that teams learn to manage and train themselves. This chapter describes a team training strategy, guided team self-correction, which helps teams to diagnose and solve their own performance problems. This is done by developing accurate and shared mental models of teamwork during structured post-performance debriefing sessions. An application of this strategy for Navy command and control teams is described and propositions for future research are offered.

Advances in Interdisciplinary Studies of Work Teams, Volume 6, pages 55-72.

ISBN: 0-7623-0655-6

INTRODUCTION

The current trend toward downsizing, in both industry and the military, has necessitated that teams learn to manage and train themselves (Smith-Jentsch, Zeisig, Acton, & McPherson, 1998). This requires team training strategies which employ inductive approaches to learning such as guided discovery learning or error-based instruction. Such strategies develop the self-regulatory mechanisms that teams need to function effectively as an autonomous unit. The purpose of this chapter is to describe a strategy for guided team self-correction of teamwork processes. Guided team self-correction involves structuring post-performance team debriefings so that team members develop accurate and shared mental models, and thus, are better able to diagnose and solve their own performance problems. The following section begins by providing an overview of the theoretical bases for our research in this area. Next, we briefly describe an application of guided team self-correction that has been used for Naval command and control teams. Finally, we identify propositions for future research in this area.

THEORETICAL BASES

Four lines of research form the bases for guided team self-correction of teamwork processes. Each of these are discussed in the following sections. First, we will begin by discussing the literature on team-related mental models. Second, the concept of team self-correction will be described. Third, we examine the impact that team members' mental models of teamwork can have on their ability to benefit from team self-correction sessions. Fourth, the cognitive stages involved in developing mental models of teamwork will be reviewed.

Team-Related Mental Models

The term *mental model* refers to knowledge structures that humans use to cognitively represent relationships within a system (Rouse & Morris, 1986). This system may include equipment (e.g., mechanical or electronic), or it may be a system which includes people and interactions among people (i.e., an organization or a team) (Gentner & Stevens, 1983; Rumelhart & Ortony, 1977). Mental models of the task, team, and equipment are expected to guide the manner in which team members perform their tasks and interact with one another (Rouse, Cannon-Bowers, & Salas, 1992). These mental models allow team members to know what events and actions to expect, when the events will occur, and how they as teammates should respond.

Three properties of team-related mental models that have been studied are: (1) consistency/stability, (2) accuracy, and (3) agreement. Each of these properties have been hypothesized to influence how team members interpret experiences and react to situations. When team members represent their mental models reliably over time and across multiple measurement techniques, this can be taken as evidence of a somewhat stable mental model. The reliability of mental model representations should have a direct impact on the extent to which they can be used to make predictions about future behavior.

The extent to which team members hold mental models that accurately represent team-related systems has been linked to performance in a few recent studies (Pape & Rentsch, 1998; Smith-Jentsch, Kraiger, Cannon-Bowers, & Salas, 1998). For example, Smith-Jentsch, Kraiger, and colleagues (1998) found that accurate mental models of teammate strengths and weaknesses were related to teamwork behavior within air traffic control teams.

The extent to which team members hold similar mental models has been referred to as agreement, overlap, or sharedness. Shared mental models among teammates have been linked to superior team performance in a number of studies (Cannon-Bowers & Salas, 1997; Klimoski & Mohammed, 1994; Rentsch & Hall, 1994). In many cases, multiple different strategies for accomplishing a task can be effective. In such cases, shared mental models are expected to improve team performance by providing a unified focus of effort for team members. For example, Blickensderfer, Cannon-Bowers, and Salas (1997) found that the degree to which team members shared expectations regarding tasks to be performed was positively related to their overall performance.

It has been argued that team-related mental models influence the way in which individuals interpret and respond to team experiences. Furthermore, three properties of mental models were described: consistency/stability, accuracy, and agreement or sharedness. We have argued that team performance is enhanced when team members hold consistent, accurate, and shared mental models. Therefore, one of the goals of team training should be to shape team-related mental models. The following section describes a strategy for accomplishing this.

Team Self-Correction

The notion behind "team self-correction" is to take advantage of opportunities for teams to debrief themselves following a significant performance event. Effective teams tend to use these opportunities to reflect on what went well and what did not, as well as why performance may have broken down and what to do differently in the future. Through the effective exchange of observations, concerns, and ideas, team self-critique discussions can enable team members to build shared mental models and, in turn, improve their performance (Blickensderfer et al., 1997). However, while some teams have a natural tendency to engage in effective "team self-correction," other teams are unsuccessful for a variety of reasons (Blickensderfer et al., 1997; Smith-Jentsch, Payne, & Johnston, 1996; Tannenbaum, Smith-Jentsch, & Behson, 1998). First, team discussions may be unfocused and fragmented. Second, team members may not possess the diagnostic skills to

Table 1. Guidelines for Effective Team Self-Correction

Prior to a team self-correction discussion, team members should be trained to use effective feedback skills:
- Specific
- Solution-oriented
- Behavioral vice, personal or insulting
- Clear and direct
- Positive as well as negative
- Willingness to accept feedback
- Request feedback from others
- Critique self

Teams should be provided with sufficient time (30-40 mins.) and a sanctioned forum following their performance to have a team self-correction discussion.

Team self-correction sessions should begin with a recap/overview to trigger team members= recall.

Team self-correction discussions should be structured using a topical outline or model of team-related systems (e.g., task, equipment, teamwork).

Team self-correction discussions should focus on the mastery of processes (how to change behavior) rather than maximization of outcomes ("do better").

An instructor, team leader, or team member should facilitate team self-correction by: Establishing a team climate which is open and supportive of learning
- Keeping the discussion focused by following a structured outline
- Modeling effective feedback skills
- Accepting feedback and ideas from others
- Coaching team members in linking concrete examples to generalizable categories of behavior and stating their feedback constructively
- Ensuring that examples do not focus exclusively on a few teammates
- Encouraging active team member participation during discussion
- Reinforcing team member participation, especially self-critique

Team self-correction discussions should culminate in specific improvement goals for the next performance.

Source: Adapted from Blickensderfer et al., 1997; Tannenbaum et al., 1998; Smith-Jentsch, Zeisig et al., 1998.

evaluate their own performance. Third, team members may not deliver their feedback to one another in a constructive way (i.e., assertively, not aggressively). Fourth, the climate within a team may not reinforce open communication and participative decision making. Each of these four problems can prevent teams from developing shared mental models or cause them to develop a shared mental model that is inaccurate.

Researchers have offered a number of guidelines to combat these potential pitfalls (see Table 1). Some of these guidelines are designed to improve team members' abilities to convey their ideas, concerns, and suggestions in a constructive manner. For example, Blickensderfer and colleagues (1997) demonstrated that team members who received feedback skills-training prior to a series of team self-correction sessions developed greater shared mental models of task expectations.

Other guidelines are designed to focus the content of team self-correction discussions. For example, Smith-Jentsch and colleagues (1996) described an approach for structuring the post-exercise debriefings given by Navy command and control teams. The approach utilized a facilitator to guide team discussions toward appropriate topics. This notion of "guided team self-correction" involves developing a structured outline, or debriefing guide, which lists specific questions designed to stimulate discussion on a particular topic (e.g., expectations, teamwork processes, teammate-specific preferences). The outline provided should reflect an accurate model of a system which team members can use to critique their own performance. This self-critique is expected to improve diagnosis and problem solving by allowing team members to work from a shared and accurate mental model. In addition, team discussions are prevented from becoming fragmented and nonproductive.

While it is argued that guided team self-correction can be used to foster many different types of team-related mental models, this chapter will focus on mental models of teamwork processes. The following section describes the role that teamwork mental models can play in diagnosing and solving team performance problems.

Mental Models of Teamwork and Team Self-Correction

Smith-Jentsch, Zeisig, and colleagues (1998) argued that when team members adopt a common frame of reference, or "shared mental model" of effective team processes they will be better able to diagnose and correct performance problems. Moreover, this common frame of reference is likely to lead team members to agree on goals for improvement and to be committed to achieving those goals. However, since team members are each likely to have had different experiences prior to joining a team, their knowledge and opinions about the nature of teamwork may differ substantially. For example, Rentsch, Heffner, and Duffy (1994) found that individuals with greater team experience tended to represent their knowledge about teamwork more consistently, using fewer categories with more abstract labels. Thus, team members with varied backgrounds may find it difficult to collectively assess their performance. In contrast, when team members share accurate mental models of teamwork, they should be better able to: (1) uncover performance trends and diagnose deficiencies, (2) focus their practice appropriately on specific goals, and (3) generalize lessons they learn to new situations (Smith-Jentsch, Zeisig et al., 1998). The manner in which mental models of teamwork are expected to improve each of these team capabilities is described below.

Uncovering Trends and Diagnosing Deficiencies

It has been argued that team members' mental models of teamwork provide them with an understanding and a structure for working together as a team (Rentsch et al., 1994). We argue that accurate mental models of teamwork will allow team members to discern common themes that explain coordination breakdowns during complex tasks. Instead of viewing every miscommunication as an independent problem needing to be remedied individually, teams who share well-developed mental models of teamwork recognize similarities among performance instances stemming from a particular "type" of problem (e.g., failure to monitor and provide backup to others).

Focused Practice and Goal Setting

Shared, accurate mental models of teamwork should also help teams to focus their efforts on specific goals that are appropriate for them. Teams that share a rich, multi-level understanding of teamwork processes should be more likely to identify specific goals rather than general goals. For example, an emergency dispatcher with a well-developed understanding of teamwork might identify a goal as "reducing excess chatter on the communication nets" rather than a more general goal, such as "improving team communications." It has been well documented that such general "do your best" goals are far less effective than specific targeted goals (Locke, Shaw, Saari, & Latham, 1981). Moreover, when team members share a mental model of teamwork, they are more likely to agree on their goals for improvement, since they are working from a common frame of reference. Team consensus on goals for improvement is desirable, since this should foster commitment to goal achievement (Brawley, Carron, & Widmeyer, 1993).

Generalizing Lessons Learned to New Situations

Since individuals use mental models to guide their behavior in new situations, accurate mental models of teamwork should help team members to generalize lessons learned about teamwork. A well-developed mental model of teamwork is expected to lead team members to focus their attention on the mastery of processes rather than maximizing performance outcomes during training. Kozlowski, Gully, Smith, Nason, and Brown (1995) examined the impact of these two types of learning orientations. Results indicated that trainees who focused on "mastery" goals were better able to generalize what they learned to a new task than those who focused on "performance" goals. Further, Kozlowski and colleagues (1995) found evidence to suggest that increased metacognitive activity, self-efficacy, and deeper learning were responsible for the observed differences in training effectiveness.

The implications of these findings in a team setting may be that team members who possess a thorough understanding of key teamwork processes should be better able to determine when such processes are appropriate and should be more confident in their ability to apply those processes when facing a novel situation. For example, if a team member knows that an important component of teamwork involves monitoring signs of stress in others and offering backup or assistance when needed, he or she can consciously look for opportunities to apply this knowledge in a variety of novel situations.

In summary, accurate and shared mental models of team processes should be helpful to team self-correction for many reasons. First, they should enable teams to uncover team-process trends and use these to diagnose performance deficiencies. Second, accurate and shared mental models of teamwork should help teams to set specific goals for improvement. Finally, accurate and shared models of teamwork should help teams to generalize lessons learned (concerning their team process performance) to new situations. Now that we have explained how mental models of teamwork processes may impact team performance, we will turn our discussion toward the mechanisms by which individuals are expected to acquire such knowledge structures.

INFORMATION EXCHANGE	COMMUNICATION
• Utilizing all available sources of information • Passing information to the right persons without having to be asked • Providing big picture updates	• Proper phraseology • Completeness of standard reports • Brevity/Avoiding excess chatter • Clarity/Avoiding inaudible comms
SUPPORTING BEHAVIOR	**INITIATIVE/LEADERSHIP**
• Monitoring & correcting errors • Providing & requesting backup or assistance to balance workload	• Providing guidance or suggestions • Stating priorities

Figure 1. Expert Model of Teamwork

Developing Mental Models of Teamwork: Cognitive Components

We have argued that mental models of teamwork guide behavior in team settings as well as team members' retrospective interpretation of team experiences. Thus, it is important that team trainers and team leaders facilitate the development of accurate and shared mental models of teamwork among team members. Anderson (1982, 1987) described three components of knowledge acquisition that we examine here in our discussion of developing mental models of teamwork through guided team self-correction: (1) declarative knowledge, (2) knowledge compilation, and (3) procedural knowledge.

Declarative Knowledge

Declarative knowledge involves remembering facts or rules, generally through verbal rehearsal or other processes for rote memorization. The first stage in developing mental models of teamwork should involve providing team members with such declarative knowledge. This would include defining the components of teamwork so that the team has a common vocabulary with which to discuss their performance. This can be accomplished during a pre-task briefing and again at the start of a post-performance debriefing. Such declarative knowledge can then become the building blocks for knowledge compilation.

Knowledge Compilation

Knowledge compilation involves understanding relationships among facts or constructs that have been learned in the declarative stage. This higher-level understanding reduces the amount of cognitive resources required to interpret new observations. This is done by allowing an individual to retrieve knowledge directly from long-term memory in a single step rather than progressing through a series of steps in working memory. The compilation of teamwork knowledge involves establishing a multi-level organization of teamwork components (Rentsch et al., 1994). This may mean, for example, that a team member views passing information, gathering information, and providing big picture summaries to team members as three components of information exchange. A novice team member may first classify an observation as a "big picture summary" and then determine that providing such a summary is part of effective information exchange in two cognitive steps. In contrast, an experienced team member may automatically classify the same example at both levels in a single step.

In order to foster the compilation of declarative knowledge about teamwork through guided team self-correction, trainees should be provided with an advanced organizer that mirrors a multi-level model of teamwork (such as the one shown in Figure 1). This "expert model" of teamwork can then serve as the basis for discussions about the relationships among higher- and lower-order teamwork components. Key features that link subcomponents within a dimension of teamwork, as well as features that differentiate dimensions of teamwork should be made explicit to trainees using concrete examples.

Procedural Knowledge

Procedural knowledge requires generalization, discrimination, and strengthening of one's conceptual understanding by applying it in novel situations. Generalization refers to the ability to identify appropriate opportunities to apply one's knowledge or skill in a new context. Discrimination, on the other hand, involves identifying features in that context that are different and thus require modifications in response. Finally, strengthening of knowledge and skills through practice is important, because it increases the likelihood that such knowledge will be applied, or at least attempted to be applied, in the future.

Therefore, the final stage of training designed to guide self-correction of teamwork processes should involve active practice extracting concrete instances of teamwork from novel team experiences or observations and organizing them within a defined "expert" model. This can be facilitated by an instructor/leader who possesses effective feedback skills and who is knowledgeable regarding the expert model of teamwork.

Summary

We have argued that shared and accurate mental models of teamwork enable team members to better: (1) diagnose performance trends, (2) focus their practice appropriately, and (3) generalize lessons learned to novel situations. It follows that one of the goals of team training should be to facilitate the development of shared, accurate mental models of teamwork. The previous sections of this chapter described three critical components of such training which follow the cognitive stages one goes through when acquiring knowledge (Anderson, 1982, 1987). The following section describes a specific application of guided team self-correction to foster mental models of teamwork for Naval teams.

AN APPLICATION OF GUIDED TEAM SELF-CORRECTION

A specific application of guided team self-correction was developed to train Naval teams. First, an *expert model* of teamwork was defined for a particular team environment. This expert model included specific teamwork behaviors (e.g., use of proper phraseology) organized within higher order teamwork dimensions (e.g., communication) that had been identified through a team training needs analysis (Smith-Jentsch, Johnston, & Payne, 1998).

Eleven teamwork behaviors, which clustered into four teamwork dimensions, were shown to differentiate experienced and inexperienced teams (see Figure 1). This expert model of teamwork became the framework for: (1) focusing team members' attention during a pre-task briefing, (2) allowing a trainer, supervisor, or team leader to collect teamwork examples systematically during the exercise, (3) organizing these observations into a structured debriefing outline, and (4) using that outline to guide a team critique.

Team Dimensional Training

The following sections describe the process and tools developed for this application, referred to as team dimensional training (TDT).

Pre-Briefing

TDT begins with a pre-task briefing. It is during this pre-task briefing that team members are first introduced to the expert model of teamwork (see Figure 1). Higher-order teamwork dimensions (e.g., initiative/leadership) are defined by specific observable behaviors (e.g., providing guidance, stating priorities). This declarative knowledge about teamwork provides team members with a common vocabulary with which to discuss their performance. TDT facilitators outline the steps that will follow for the team, including requirements for their participation, and convey genuine interest in team-member input and growth.

Organization of teamwork components around higher-order dimensions provides team members with an advanced organizer which supports knowledge compilation. A TDT pre-briefing guide provides definitions and structure to aid facilitators (see Smith-Jentsch, Zeisig et al., 1998). These facilitators can be team trainers, team leaders, or supervisors.

Team Performance Event

TDT can be used in conjunction with either a team training exercise or a critical performance event in the actual work environment. This critical event could be an important presentation for a marketing group or a play-off game for a basketball team. During the performance event, a TDT facilitator (e.g., instructor, supervisor, team leader) seeks out and records examples which exemplify positive and negative instances of teamwork behaviors within the expert model. Following the performance event, the facilitator organizes his or her notes within the TDT debriefing guide, which follows an outline that mirrors the expert model of teamwork. This provides a structure for the debriefing that flows dimension by dimension with component behaviors organized accordingly (see Figure 2). TDT facilitators record 1-2 examples of each component behavior within the outline provided. In our experience, we have found that a well-prepared 30-minute team

The third teamwork dimension is SUPPORTING BEHAVIOR. One component of supporting behavior is error correction. This involves monitoring for team errors, bringing an error to the team's attention, and seeking that it is corrected.

Question	Example
Give me an example of an error that your team caught and corrected. - How was it corrected?	+ TIC told AAWC she had transposed numbers when reporting a track.
In retrospect, what errors were not caught and corrected that could have been? - How could these errors have been caught and corrected, and by whom?	- TAS Operator covered track 2510 with birds- but SWC ordered track 2501. No one caught or corrected, and wrong track was engaged.

The Second component of supporting behavior is providing backup/assistance. This involves noticing that another team member is overloaded or having difficulty performing a task and providing assistance to them by actually taking on some of their workload.

Question	Example
Give me an example of when assistance or "backup" was provided to reduce another team member's workload. - How did this improve the team's ability to deal with key events in the exercise?	+ Track Sup made identifications for ID operator while he was trouble-shooting the link
Describe for me an instance when someone on the team could have benefited from backup that was not provided. - What kind of backup could have been given, and by whom?	- Channel alpha suffered a casualty and channel bravo did not transmit channel alpha's information

Figure 2. Sample Page from a TDT Debriefing Guide

training exercise can provide more than enough examples to be discussed in a TDT debriefing (Smith-Jentsch, Zeisig et al., 1998). The concrete examples of teamwork processes demonstrated during the performance event will later serve as a springboard for team discussion. These examples can be used to convey distinctions among behavioral categories and to convince team members that teamwork does, in fact, contribute to their success or failure as a team.

Debriefing

A TDT debriefing begins by first recapping the timeline of a performance event. This helps to refresh team members' memories and to clarify what happened *before* getting into why it happened. Next, the TDT facilitator guides the team in a self-critique of their teamwork by using probes listed in the debriefing guide. These probes are open-ended questions about each of the component behaviors within the expert model of teamwork (e.g., "Give me an example of a time when backup was provided that really helped the team."). This requires team members to critically review the team's performance and identify a concrete example which appropriately fits the category in question (e.g., positive backup behavior). The process of linking concrete examples of teamwork to more abstract categories within an expert model of teamwork is intended to develop team members' procedural knowledge. Such knowledge is expected to be critical for generalizing lessons learned to new performance situations.

Facilitators reserve their own observations for instances when the team member input is not forthcoming or to clarify a distinction between behavioral categories. In addition, a TDT facilitator coaches team members in stating their feedback to one another in a constructive manner. Follow-up questions require team members to identify solutions to problems noted (e.g., "Who could have provided you backup, and what kind of backup could they have provided?"). Finally, the TDT facilitator guides team members in formulating specific, attainable, solution-oriented goals. One goal is selected under each of the higher-order dimensions.

Four Basic Tenets of TDT

Preliminary data indicated that, after TDT, trainees' mental models of teamwork were in greater agreement and that these mental models were also more accurate (i.e., closer to the data-driven expert model) (Smith-Jentsch, Campbell, Ricci, & Harrison, 1998). We have experimented with TDT in numerous team environments within the Navy (e.g., combat systems, damage control, engineering, aviation, seamanship/navigation), and recently with control teams at a nuclear power plant. Implementation of TDT in each of these environments has varied somewhat. For example, team size, performance duration, physical distribution of team members, and type of facilitator (e.g., leader, instructor) are just a few variables that differed across TDT evolutions. We have found that TDT was flexible enough to support training in each of these environments. Thus, we believe that the TDT approach would be useful for many industry teams (manufacturing, medical, police, software development, and others). The caveat is that four basic tenets must be upheld.

Tenet 1. A team is provided with both the time and sanctioned forum to discuss teamwork issues relatively soon after performance in an exercise or critical workplace event.

Guided team self-correction should occur as soon as possible following a performance event. The longer the lag between the end of the performance session and the team discussion, the less useful the team discussion. This is because (1) human memory is fallible and teams may simply forget what occurred during their performance, and (2) feedback tends to be more effective immediately following performance (Ilgen, Fisher, & Taylor, 1979). Furthermore, the most effective TDT debriefings tend to last 30-40 minutes (Smith-Jentsch, Zeisig et al., 1998). If 30 minutes cannot be spared for a critique of teamwork processes, team members tend to doubt that their input is truly desired and to feel too rushed to participate fully. Moreover, issues raised and not resolved may do harm to team cohesion. For TDT to be effective, team members must be convinced that teamwork processes are important for their performance. At the very least, an organization must first demonstrate its belief that teamwork is a priority by providing the time and a sanctioned forum specifically devoted to diagnosing and solving teamwork problems.

Tenet 2. An accurate model of domain-specific teamwork processes is used to guide the team self-correction discussion.

As we have argued earlier, team members may each come to a team with different ideas about what constitutes effective teamwork. Without a common frame of reference regarding teamwork, a team may find it difficult to agree on a diagnosis of their own deficiencies or solution to their problems. In this way, unguided team self-correction can become fragmented or unfocused leaving team members frustrated and potentially hostile toward one another. TDT provides team members with a common frame of reference (i.e., mental model) and a vocabulary with which to discuss their teamwork processes. Using the debriefing guide as an outline, the facilitator keeps the team discussion focused. Behavioral instances are examined for their learning value in terms of the teamwork model. In this way, team members are better able to uncover performance trends, diagnose deficiencies, and generalize lessons learned to new situations.

The TDT methodology is expected to generalize to any team environment. The model of teamwork embedded in TDT thus far was found to be generalizable to multiple team environments within the Navy (i.e., aviation, seamanship, engineering, damage control, combat systems). However, when implementing the TDT methodology in a new team environment (e.g., project action teams), some sort of team training needs analysis should be performed to examine the generalizability of the specific dimensions and component behaviors within this teamwork model. Ultimately, the topical outline for TDT debriefings should reflect teamwork processes that lead to superior performance in a desired team setting.

Tenet 3. A facilitator is used to (a) keep the team's discussion focused, (b) establish a positive climate, (c) encourage and reinforce active participation, (d) model effective feedback skills, and (e) coach team members in stating their feedback in a constructive manner.

For TDT to be effective, a facilitator is needed to guide the team discussion. This facilitator can be a supervisor, team leader, or team trainer/instructor. In order to be effective, a TDT facilitator should be proficient in categorizing behavioral examples within the expert model of teamwork and should possess effective feedback skills. The climate established by the facilitator has a significant impact on team members' willingness to offer observations and to accept constructive criticism from others. TDT facilitators are trained to draw out comments from less outspoken team members, as well as to restate input that is delivered in a nonconstructive manner.

Tenet 4. Goals for improvement on specific teamwork behaviors within the expert model are identified.

TDT incorporates goal setting to focus team members' efforts and to motivate them to improve on a limited set of specific, solution-oriented goals under each teamwork dimension. Goal-setting combined with feedback has been directly related to performance improvement (Kanfer, 1990). Performance improvement on these goals should be tracked and incorporated into subsequent team prebriefings and debriefings.

Summary

We believe the TDT methodology will be useful for varied types and sizes of teams in industry, however, four basic tenets must be upheld. First, the team must be provided the time and forum to discuss teamwork issues following their performance. Second, a model of teamwork should provide the structure to the discussions. Third, a facilitator (an instructor, team leader, or team member) should keep the discussion focused, maintain a learning climate, encourage participation, and model effective discussion skills during the self-correction session. Finally, the team should set improvement goals which correspond to the expert model which guides the discussion.

This chapter has discussed the conceptual underpinnings of using structured team debriefing sessions to improve self-regulation of teamwork processes. We described a team training strategy designed to produce shared mental models of teamwork and to improve diagnosis and transfer of lessons learned from team performance events. We now turn to propositions for future research in this area.

PROPOSITIONS FOR GUIDED TEAM SELF-CORRECTION

In light of the above discussion, a number of propositions can be derived about guided team self-correction and shared mental models of teamwork.

Proposition 1. Team members who share accurate mental models of teamwork will more accurately diagnose performance problems.

Accurate mental models of effective teamwork are expected to aid team members in determining which processes are working effectively and which are not. To the extent that teammates agree on the components of teamwork, they should be better able to discern trends and avoid misunderstandings.

Proposition 2. Team members who share mental models of teamwork will report greater commitment to established teamwork-related goals.

When team members share mental models of what effective teamwork looks like, they are more likely to agree on specific goals for improvement. This should foster increased commitment to previously set goals, since team members are unified in their focus.

Proposition 3. Team members who share mental models of teamwork will report greater collective efficacy.

It is expected that when team members share a common mental model about teamwork they will be better able to self-regulate. Team members are likely to recognize this fact and therefore gain greater confidence that their team can adapt to changing demands and maintain stable performance.

Proposition 4. Team members who share accurate mental models of teamwork will be more effective at generalizing lessons learned to novel situations.

Mental models are believed to guide people's interactions with the environment. Thus, team members who can link concrete experiences to more abstract constructs within a model of teamwork should be better able to identify opportunities to generalize lessons learned to new contexts. However, if team members do not share mental models, they may interpret lessons learned differently.

Proposition 5. Guided team self-correction structured around an expert model of teamwork will help teams to build accurate and shared mental models of effective team processes.

Repeated experience at linking concrete examples of performance to more general categories during guided team self-correction sessions has been shown to produce more accurate and shared mental models of teamwork for Navy command and control teams. It is expected that this technique will be effective across a variety of team settings.

Proposition 6. Guided team self-correction can be used to develop many types of mental models (e.g., teamwork processes, task expectations, task strategy, equipment).

While this chapter has focused on teamwork processes, the methodology described could be used to develop other types of mental models that are necessary for effective team performance. For example, an expert model of how equipment is operated could be used to structure team self-correction discussions.

CONCLUSION

In sum, guided team self-correction appears to be a viable method of helping team members to help themselves. This chapter discussed the use of this strategy for developing teamwork mental models. A specific application, team dimensional training (TDT), has proven to be effective, practical, and generalizable to a variety of Navy team environments. Although further testing is needed, preliminary evidence has indicated that team members' mental models of teamwork become more similar to one another and closer to the expert model after using TDT (Smith-Jentsch, Campbell et al. 1998). Additionally, instructors who have used this technique in operational settings report that it accelerates the rate with which teams achieve training objectives (Smith-Jentsch, Zeisig et al., 1998).

NOTE

The views expressed herein are those of the authors and do not necessarily represent those of the organizations with which they are affiliated.

REFERENCES

Anderson, J.R. (1982). Acquisition of cognitive skill. *Psychological Review, 89*(4), 369-406.

Anderson, J.R. (1987). Skill acquisition: Compilation of weak method problems. *Psychological Review, 94,* 192-210.

Blickensderfer, E., Cannon-Bowers, J.A., & Salas, E. (1997). *Training teams to self-correct: An empirical investigation.* Paper presented at the 12th annual meeting of the Society for Industrial and Organizational Psychology, St. Louis, MO.

Brawley, L.R., Carron, A.V., & Widmeyer, W.N. (1993). The influence of the group and its cohesiveness on perceptions of group goal-related variables. *Journal of Sport and Exercise Psychology, 15*(3), 245-260.

Cannon-Bowers, J.A., & Salas, E. (1997). A framework for developing team performance measures in training. In M.T. Brannick, E. Salas, & C. Prince (Eds.), *Team performance assessment and measurement: Theory, research, and applications* (pp. 45-62). Hillsdale, NJ: LEA.

Gentner, D., & Stevens, A.L. (Eds.). (1983). *Mental models.* Hillsdale, NJ: LEA.

Ilgen, D.R., Fisher, C.D., & Taylor, M.S. (1979). Consequences of individual feedback on behavior in organizations. *Journal of Applied Psychology, 64,* 349-371.

Kanfer, R. (1990). Motivation theory and industrial and organizational psychology. In M.D. Dunnette & L.M. Hough (Eds.), *Handbook of industrial and organizational psychology* (pp. 75-170). Palo Alto, CA: Consulting Psychologists Press, Inc.

Klimoski, R., & Mohammed, S. (1994). Team mental model: Construct or metaphor? *Journal of Management, 20,* 403-437.

Kozlowski, S.W.J., Gully, S.M., Smith, E.A., Nason, E.R., & Brown, K.G. (1995). Sequenced mastery training and advance organizers: Effects on learning, self-efficacy, performance, and generalization. In R.J. Klimoski (Chair), *Thinking and feeling while doing: Understanding the learner in the learning process.* Symposium conducted at the 10th annual conference of the Society for Industrial and Organizational Psychology, Orlando, FL.

Locke, E.A., Shaw, K.N., Saari, L.M., & Latham, G.P. (1981). Goal setting and task performance: 1969-1980. *Psychological Bulletin, 90,* 125-152.

Pape, L.J., & Rentsch, J.R. (1998). The effects of trust and perspective-taking on team members' schema similarity. In K.A. Smith-Jentsch (Chair), *To be a team is to think like a team.* Symposium presented at the 13th annual conference of the Society for Industrial and Organizational Psychology, Dallas, TX.

Rentsch, J.R., & Hall, R.J. (1994). Members of great teams think alike: A model of team effectiveness and schema similarity among team members. In M.M. Beyerlein & D.A. Johnson (Eds.), *Advances in interdisciplinary studies of work teams: Vol. 1. Theories of self-managing teams* (pp. 223-261). Greenwich, CT: JAI Press.

Rentsch, J.R., Heffner, T.S., & Duffy, L.T. (1994). What you know is what you get from experience: Team experience related to teamwork schemas. *Group and Organization Management 19*(4), 450-474.

Rouse, W.B., Cannon-Bowers, J.A., & Salas, E. (1992). The role of mental models in team performance in complex systems. *IEEE Transactions on Systems, Man, and Cybernetics, 22*(6), 1296-1308.

Rouse, W.B., & Morris, N.M. (1986). On looking into the black box: Prospects and limits in the search for mental models. *Psychological Bulletin, 100,* 349-363.

Rumelhart, D.D., & Ortony, A. (1977). The representation of knowledge in memory. In R.C. Anderson & R.J. Spiro (Eds.), *Schooling and the acquisition of knowledge* (pp. 99-135). Hillsdale, NJ: LEA.

Smith-Jentsch. K.A., Campbell, G.E., Ricci, K., & Harrison, J.R. (1998). Training mental models of teamwork through structured post-exercise debriefs. In K.A. Smith-Jentsch (Chair), *To be a team is to think like a team.* Symposium presented at the 13th annual conference of the Society for Industrial and Organizational Psychology, Dallas, TX.

Smith-Jentsch, K.A., Johnston, J.H., & Payne, S.C. (1998). Measuring team related expertise in complex environments. In J.A. Cannon-Bowers & E. Salas (Eds.), *Making decisions under stress: Implications for individual and team training* (pp. 61-87). Washington, DC: American Psychological Association.

Smith-Jentsch, K.A., Kraiger, K., Cannon-Bowers, J.A., & Salas, E. (1998). A data-driven model of precursors of teamwork. In K. Kraiger (Chair), *Team effectiveness as a product of individual, team, and situational factors.* Symposium presented at the 13th annual conference of the Society for Industrial and Organizational Psychology, Dallas, TX.

Smith-Jentsch, K.A., Payne, S.C., & Johnston, J.H. (1996). Guided team self-correction: A methodology for enhancing experiential team training. In K.A. Smith-Jentsch (Chair), *When, how, and why does practice make perfect?* Paper presented at the 11th annual conference of the Society for Industrial and Organizational Psychology, San Diego, CA.

Smith-Jentsch, K.A., Zeisig, R.L., Acton, B., & McPherson, J.A. (1998). Team dimensional training. In J.A. Cannon-Bowers & E. Salas (Eds.), *Making decisions under stress: Implications for individual and team training* (pp. 271-297). Washington, DC: American Psychological Association Press.

Tannenbaum, S.I., Smith-Jentsch, K.A., & Behson, S. (1998). Training team leaders to facilitate team learning and performance. In J.A. Cannon-Bowers & E. Salas (Eds.), *Making decisions under stress: Implications for individual and team training* (pp. 247-270). Washington, DC: American Psychological Association Press.

INFLUENCE IN SELF-MANAGING TEAMS
A CONCEPTUAL EXAMINATION OF THE EFFECTS OF CONTENT AND CONTEXT FACTORS ON TRAINING ENROLLMENT DECISIONS

Wanda J. Smith, L. Scott Casino, and
Christopher P. Neck

ABSTRACT

Organizations are increasingly relying on a self-managing team's (SMTs) structure to maximize their performance. Yet, little is known about how SMT members interact. This chapter seeks to conceptually enrich our understanding of the ways in which SMT members attempt to influence their team members when training enrollment decisions are being made. A contingency model of influence in the SMT context is introduced. It is postulated that a SMT member's choice of influence tactics is dependent on three situational factors: influence agent's base of power, the type of team cohesion, and the content of the training. The model is used to specify more precisely which influence tactics will be used in SMT training enrollment decisions.

Advances in Interdisciplinary Studies of Work Teams, Volume 6, pages 73-90.

ISBN: 0-7623-0655-6

INTRODUCTION

Organizational members, especially members of SMTs, find it increasingly necessary to influence colleagues to accomplish work objectives, acquire valued rewards, and to achieve desired career goals (Cohen & Bradford, 1989). One career goal currently receiving attention is remaining "employable" through continuous learning (Albrecht, 1996; Koonce, 1996). As a result, employees have increased requests for training in order to maintain or improve their knowledge base (Brousseau, Driver, Eneroth, & Larsson, 1996).

Traditionally, managers have made final decisions regarding who attends what type of training, when, and for how long. Yet, in self-managing teams—comprised of interdependent individuals who self-regulate their behavior on relatively whole tasks (Cohen, Ledford, & Spreitzer, 1996; Goodman, Devadas, & Hughson, 1988), these developmental decisions are often made by the team and facilitated by a manager (Moorhead, Neck, & West, 1996; Orsburn, Moran, Musselwhite, Zenger, & Perrin, 1990; Zenger, Musselwhite, Hurson, & Perrin, 1994). Given that employees in SMTs are more responsible for seeing to the teams' own training needs, team members must be able to convince their colleagues of the utility of additional training. In short, SMT members need to be able to convince fellow team members that the benefits of such training exceed the costs to the team (i.e., a particular team member's absence from the team during the training period). Also assuming that the utility of such training is agreed upon, members need to determine who will attend such training and when such training will occur. Accordingly, the widespread implementation of work teams coupled with the increased responsibility of employees to see to their own development, raises an interesting question: "what influence tactics will SMT members use to influence decision as regarding *who* attends *what* type of training?"

We believe exploring the answer to this question is a fruitful exercise given the contention that training is a very important element necessary for long-term SMT effectiveness (Manz & Sims, 1993). In brief, given the importance of training to SMT success, it seems logical that choices within the team relating to the type of training chosen and who will receive the training are indeed critical decisions. This logic stems from the argument that without the appropriate skills necessary to succeed, SMTs will function in an ineffective manner (Manz, Neck, Mancuso, & Manz, 1997). Accordingly, training decisions to acquire such skills necessary for SMT success are crucial. Thus, a greater understanding of the factors that lead to and/or influence such decisions would take us one step closer to understanding the within-team processes that impact SMT effectiveness.

The social influence literature has, for the most part, focused on identifying the various contexts in which different types of influence tactics are used more frequently (Erez, Rim, & Keider, 1986; Kipnis, Schmidt, & Wilkinson, 1980); in what direction (Barry & Bateman, 1992; Yukl & Falbe, 1990; Yukl & Tracey, 1992); and for which objectives (Ansari & Kapoor, 1987; Hinkin & Schriesheim,

Levels					Contingencies
I	Consultative		Coercive		Influence Tactics
II	Task-Oriented or Team-Oriented	Task-Oriented or Team-Oriented	Team-Oriented	Team-Oriented	Nature of Training
III	Personal		Reciprocity		Sources of Power
IV	High Social	Low Social	Low Task	High Task	Type of Team Cohesion

Figure 1. A contingency Model of Influence Choices in an SMT Context

1990; Schmidt & Kipnis, 1984). Other scholars have examined which combination (Yukl, Falbe, & Youn, 1993) and sequencing (Kipnis & Schmidt, 1988) are most effective.

This literature has largely ignored the role of influence tactics in self-managed teams and the effects of team dynamics on the influence process. This chapter seeks to conceptually enrich our understanding of intra-team social influence processes, particularly the strategies which SMT members utilize to influence their team when making training enrollment decisions. We do so by targeting three contingency variables and predicting their impact of the influence agent's strategy selection.

Given the widespread use of contingency approaches to leadership (c.f., Fiedler, 1964; House, 1971), it makes sense that such an orientation holds promise for understanding social influence. In fact, research has shown that an agent's choice of an influence tactic takes into account the reaction they anticipate from the target (Ansari & Kapoor, 1987). These findings indicate that an agent attempting to influence their superiors used upward appeals and ingratiation when they believed their superiors were inclined to be highly authoritarian. In contrast, subordinates reported using rational persuasion when they believed their superiors were highly participative. One interpretation of these findings is that a SMT member's choice of influence strategies are affected, in part, not simply by his or her own characteristics, but also situational factors such as his or her beliefs about the likely effects of his or her actions. Along these lines, the basic premise of this chapter is that an agent's choice of influence tactics is *contingent* upon three situational factors: the agent's source of power, the nature of the relationship among SMT members, and the nature of the training request. A contingency model of influence choices in a SMT context is presented in Figure 1.

Interpretation of the model is as follows. In the first level, Yukl's (1994) eight general influence tactics are included in two categories: cooperative and coercive. In the remaining levels of the model, three contingency factors are identified as determinants of which influence strategy will likely be used by a SMT member. The content of the influence attempt is illustrated in Level II. Specifically, is the agent requesting training designed to build interpersonal team skills (team oriented) or to gain technical skills important to the SMT's ability to complete its tasks (task-oriented). In Level III, the impact of the agent's source of power is depicted. Generally, SMT members possess power derived from two sources: (1) personal characteristics (expert or referent), and (2) reciprocal alliances (social networks centered around mutual gain). The final factor, the nature of the relationship within the SMT, refers to the extent to which the SMT is task or socially cohesive (Level IV). That is, are team members attracted to the team primarily for task-related reasons or for relationship-related reasons?

In our model in Figure 1, we contend that when requesting either task- or team-oriented types of training, SMT agents will select to use more cooperative influence tactics if they have personal sources of power and the SMT is socially cohesive. In contrast, we believe that SMT agents will use more coercive influence tactics when they are relying on reciprocal alliances, the SMT's cohesion is centered around its task, and the agent is requesting to attend team-oriented training.

Before we examine the theoretical justifications for the linkages reflected in the above model, we will discuss the singular impact of each of the three situational variables on influence choices among traditional hierarchical relationships, followed by an analysis of how each choice pattern may be reflected in a SMT. Accordingly, we will discuss how three sources of power available to SMT members (referent, expert, and reciprocal alliances) will affect their choice of influence tactics. Next, we will distinguish between two types of cohesion and analyze how each may affect the choice of influence tactics, followed by a discussion indicating the effects of the nature of training on the tactics used. Finally, the combined interaction effects of these variables will be revisited.

Since we have elected to limit the scope of this chapter to self-managed work teams, we will review the characteristics that distinguish SMTs from other work teams.

SELF-MANAGING WORK TEAMS: A DEFINITION

Self-managing teams are receiving increased emphasis in both academic and practitioner literatures (e.g., Cohen & Ledford, 1994; Hackman, 1986; Katzenbach & Smith, 1993; Lawler, 1986, 1993; Manz & Sims, 1987, 1993; Orsburn et al., 1990; Walton, 1985). The theoretical foundation for SMTs has been derived primarily from socio-technical systems theory. Socio-technical systems theory prescribes joint optimization of the social and technical aspects of work (Cummings, 1978; Emery & Trist, 1969; Susman, 1976). This optimization frequently results in the adoption of self-managing teams because "a group can more effectively allocate its resources when and where required to deal with its total variance in work conditions than can an aggregate of individuals each of whom is assigned a portion of the variance" (Susman, 1976, p. 183).

SMTs can be described as groups of interdependent individuals that can self-regulate their behavior on relatively whole tasks (Cohen, Ledford, & Spreitzer, 1996; Cummings & Griggs, 1977; Goodman et al., 1988; Hackman, 1987; Manz, 1992). Various researchers (e.g., Cohen et al., 1996; Hackman, 1987; Manz & Sims, 1993; Moorhead et al., 1996; Orsburn et al., 1990; Zenger et al., 1994) have described differentiating characteristics of SMTs—that is, characteristics that distinguish SMTs from other forms of traditional teams (e.g., intra-functional teams, problem-solving teams, cross-functional teams).

These distinguishing SMT characteristics include: (1) *task assignment*: employees perform interdependent tasks, and are mutually responsible for making a product or providing a service; (2) *decision making autonomy*: employees have discretion over decisions traditionally made by management, such as determining work assignments and schedules, resolving team productivity and interpersonal problems, assessing team training needs (e.g., what type of training is needed, who will receive this training, etc.), and conducting team meetings; employees also establish their own goals and objectives, provide performance feedback, and make necessary corrections; (3) *skill requirements*: members possess a variety of skills necessary to complete the task, produce the product or perform the service and thus limit dependence on external resources for task performance; (4) *compensation and performance feedback*: employees are usually compensated for the skills (number of tasks) they perform; and team output and performance feedback (as opposed to individual output and individual performance feedback) is rewarded for the group as a whole; (5) *supervision of the group*: managers of "self-managing teams," due to the nature of these forms of work groups, are "facilitators" as opposed to hierarchical "top-down" primary decision makers.

A plethora of self-managing team applications have occurred in both the manufacturing and the service sector. Numerous case studies have been reported, including a dog-food plant (Walton, 1977), a financial investment firm (Sims, Manz, & Bateman, 1993), a paint manufacturing plant (Poza & Markus, 1980), small parts manufacturing (Manz & Sims, 1987), an airline (Cohen & Denison, 1990), coal mines (Trist, Susman, & Brown, 1977), an independent insurance firm (Manz & Angle, 1986), a mental health hospital (Shaw, 1990), a warehouse (Manz, Keating, & Donnellon, 1990), a paper mill (Manz & Newstrom, 1990), and a telecommunications company (Cohen & Ledford, 1994).

Positive outcomes that have been attributed to the implementation of self-managing teams include increased productivity, quality, employee satisfaction, and improved quality of work life for employees, as well as decreases in absenteeism and turnover (Cohen & Ledford, 1994; Corderey, Mueller, & Smith, 1991; Herbst, 1962; Lawler, 1986; Manz & Sims, 1987; Saporito, 1986; Verespej, 1990). SMTs and a socio-technical systems perspective have also been advocated as a means for improving organizational flexibility and thereby promoting adaptability to a rapidly changing business environment (Manz & Stewart, 1997). Further research is needed that clarifies the benefits of adopting a team-based organizational structure. The general conclusion, however, is that empowered teams improve the effectiveness of organizations (U.S. Department of Labor, 1993).

Given these many benefits that have been attributed to the implementation of self-managing teams, research needs to be executed that addresses our understanding of within-team processes that impact the effectiveness (e.g., performance) of teams. For example, a better understanding of which influence strategies are most popular is the first step in identifying which influence tactics will be most effective, each of which will indirectly impact team performance.

The next section of this chapter discusses the social influence process, opening with the definition and clarification of a number of terms. In addition, we briefly review research about social influence in organizations.

SOCIAL INFLUENCE PROCESSES IN ORGANIZATIONS

Management scholars now recognize that power and influence are cornerstones of our understanding of organizational behavior (Mintzberg, 1983; Pfeffer, 1981). Yet, the study of power and influence has been largely hampered by operational or conceptual ambiguity. The extent to which these concepts are interrelated is illustrated in how commonly many theorists use the word "influence" to describe power (Pfeffer, 1992).

In this chapter, we define *power* as the capability to get someone to do something. In contrast, social *influence* is the application of power and refers to the behaviors or actions of one party (the "agent") which are performed to affect a second party's (the "target") attitudes, perceptions, or behaviors (Yukl, 1989). This subtle distinction between power and influence centers around the concepts of potential versus action. Power is the potential to influence and is often described as a set of characteristics, while influence is power in action and deals with agent behaviors.

Table 1. Political Tactics Derived from Research

Consultation Tactics	Seeking your participation in making a decision or planning how to implement a proposed policy, strategy, or change.
Rational Persuasion	Using logical arguments and factual evidence to persuade you that a proposal or request is viable and likely to result in the attainment of task objectives.
Inspirational Appeals	Making an emotional request or proposal that arouses enthusiasm by appealing to your values and ideals or by increasing your confidence that you can do it.
Ingratiating Tactics	Seeking to get you in a good mood or to think favorably of the influence agent before asking you to do something.
Coalition Tactics	Seeking the aid of others to persuade you to do something, or using the support of others as an argument for you to agree also.
Pressure Tactics	Using demands, threats, or intimidation to convince you to comply with a request or to support a proposal.
Upward Appeals	Persuading you that the request is approved by higher management, or appealing to higher management for assistance in gaining your compliance with the request.
Exchange Tactics	Making explicit or implicit promises that you will receive rewards or tangible benefits if you comply with a request or support a proposal, or reminding you of a prior favor to be reciprocated.

Source: Adapted from Yukl, G., & Falbe, C.M. (1990). Influence tactics and objectives in upward, downward, and lateral influence attempts. *Journal of Applied Psychology, 75*, 133.

Table 2. Frequency of Use of Political Influence Tactics

Tactic	*Rank Order of Downward Use*	*Rank Order of Lateral Use*	*Rank Order of Upward Use*
Consultation Tactics	1	1	2
Rational Persuasion	2	2	1
Inspirational Appeals	3	3	3
Ingratiating Tactics	4	4	5
Coalition Tactics	5	5	4
Pressure Tactics	6	7	7
Upward Appeals	7	6	6
Exchange Tactics	8	8	8

Source: Adapted from Yukl, G., & Falbe, C.M. (1990). Influence tactics and objectives in upward, downward, and lateral influence attempts. *Journal of Applied Psychology, 75*, 139.

A brief overview of the influence literature follows. Surprisingly, no studies were found examining the antecedents and outcomes of influence processes in SMT settings. Most of the influence literature has focused on managerial choice and effectiveness when seeking to influence peers, subordinates, and supervisors. This literature has also concentrated on the impact of agent's power and the content of the agent's request on agent-influence choices. Using this general influence literature as a frame of reference, this chapter will predict a pattern of influence choices likely made by SMT agents. We now turn our attention to how the social influence literature evolved.

In 1980, David Kipnis and colleagues initiated a stream of research investigating how people influence each other in organizations. Specifically, these researchers asked individuals to report how they get others to do what they want them to do (Kipnis et al., 1980). Initial research identified over three hundred influence tactics (Hinkin & Schriesheim, 1990). However, statistical refinements and replications by other researchers over a 10-year period has revealed eight general influence tactics (Barry & Bateman, 1992; Erez et al., 1986; Kipnis et al., 1980; Yukl & Falbe, 1990, 1991; Yukl & Tracey, 1992).

In addition to identifying types of influence tactics, researchers were also interested in determining how often individuals used the tactics. Some tactics were found be to used more frequently than others—Table 1 presents a list of the eight general tactics in descending order of preference. Generally speaking, more coercive tactics are used less frequently than cooperative tactics. More recent research has focused on whether agents use different tactics with different targets such as peers, superiors, and subordinates (Yukl & Falbe, 1990). Results, summarized in Table 2, show that, regardless of the target's hierarchical status relative to the agent, preferences for type of influence strategy used remained fairly consistent.

We believe this ranking of consultative and coercive influence tactics will also be reflected in the SMT environment, and is included in Level I of the contingency model. Attempts to specify in more precise terms which influence tactic will be used and when, requires further analyses of the impact of several situational factors.

Before we examine the effects of the contingencies included in the model, we will review the characteristics that distinguish two types of training. Next, we will suggest which influence tactics will be used given the nature of the training request. This matching is reflected in Level II of the contingency model (Figure 1).

THE NATURE OF TRAINING IN THE SELF-MANAGING TEAM ENVIRONMENT

Most theories of work team performance include, as an important influence on team effectiveness, the presence of a viable training system in the organizational context which ensures that team member knowledge, skills, and abilities are sufficient for meeting team task requirements (Campion, Medsker, & Higgs, 1993; Cohen et al., 1996; Hackman, 1987). Training systems typically include training needs assessment, training and development, and evaluation components which are designed to facilitate the systematic development of job-related knowledge, skills, and abilities (KSAs). In the team environment, training can be classified in two broad categories: training can be either *task-oriented*, involving the development of task-related skills, or *teamwork-oriented*, which concerns the development of behaviors related to fulfilling social/interpersonal requirements of team functioning (Salas, Dickinson, Converse, & Tannenbaum, 1992; Tannenbaum & Yukl, 1992).

In a traditional hierarchical organization, training decisions are mostly determined by organizational, situational, and task needs. As mentioned previously, the initial component in a training system is needs assessment, which involves organizational analysis, task and KSA analysis, and person analysis (Goldstein & Gilliam, 1990). Organizational analysis provides information about the efficacy of using training to meet organizational human resource needs and about potential barriers to the viability of training by examining the macro-level variables which affect training programs such as organizational goals, training climate, and organizational constraints. The second step in needs assessment, task and KSA analysis, determines the tasks which comprise a job and the knowledge, skills, and abilities relevant to performing those tasks. Finally, person analysis is an assessment of trainee capabilities in terms of determining the absence, in trainees, of critical KSAs identified in the previous step. Training traditionally falls under the purview of the human resources function which administrates and conducts the three stages of the training process (i.e., needs assessment, the actual training program, and training evaluation). The HR function may utilize the inputs of both line managers and potential trainees in making training decisions; this input is often limited to identifying skills deficiencies and assessing trainee preferences regarding program content (Baldwin, Magjuka, & Loher, 1991).

However, literature on the future organizational forms and organizational change suggests that, given the tremendously turbulent environments that current and future organizations will face, individual organizational members must shoulder more responsibility for their own development (Lawler, 1993; Mohrman & Mohrman, 1993). It is suggested that the nature of employer-employee relations is shifting towards a transactional form, that is, one where individuals are employed based on the value of their knowledge and skills to the firm (Mirvis & Hall, 1994). We believe this is particularly pertinent in the team environment where self-managed teams are given autonomy and power to make decisions which traditionally were the duties of department heads and supervisors. This decision-making ability gives a team the power to handle work problems and increase effectiveness through a number of mechanisms, such as improving team processes and enhancing team capabilities, and includes the capability to determine the training and development of individual members.

In the SMT environment, little attention has been directed to the role of the training function—other than to note the importance of training and to debate the content of the training. There is little guidance as to how to design team training systems (Cannon-Bowers, Tannenbaum, Salas, & Volpe, 1995). By design, a self-managed team is responsible for determining their own training needs (Moorhead et al., 1996; Orsburn et al., 1990; Zenger et al., 1994) which suggests that a team engages in an assessment process similar to that in a traditional organization. This process involves analyzing team task requirements from performance objectives, utilizing performance feedback to identify deficiencies in team members' current competencies and deriving a training strategy to fill these deficiencies. Cianni and Wnuck (1997) discuss a model of team-based career systems and argue teams are an integral element of employee development by providing skills feedback, identifying growth opportunities, mentoring, and offering training opportunities.

But what about the role of individual team members in this process? Each team member shares responsibility for performing the activities identified by Cianni and Wnuck and can use this performance monitoring information to help determine developmental needs. We posit that, as individual team members become aware of a developmental need for their team, they can select from a variety of influence tactics to convince their peers that training is needed. With respect to this chapter, we focus our analysis on a team member's attempt to convince their team that training is needed and that they should be the one to receive the training.

As noted previously, team training may be either task-related (sometimes referred to as technical training) or teamwork-related ("soft") training. We propose that the nature of training being sought by a particular team member indirectly impacts the type of influence tactic employed. In Figure 1, we propose that when an SMT member is requesting to attend team-oriented and task-oriented training, a range of cooperative tactics will likely be used (See Table 2 for the top four cooperative tactics). The model also suggests that more coercive influence strategies may be used to request team-oriented training under a particular set of social conditions. One of the critical contingencies is the influence agent's base of power. A review of the literature follows.

POWER IN ORGANIZATIONS

More than 30 years ago, French and Raven identified bases of power derived from two sources: (1) the characteristics individuals possess and (2) the nature of the relationship between individuals. They proposed that power arises from five different bases: reward power, coercive power, legitimate power, expert power, and referent power (French & Raven, 1959). Reward and coercive power are demonstrated when a person has control over desired rewards or the ability to punish, respectively. Legitimate power is derived directly from a person's title and position in the organizational hierarchy. Expert power originates when a person is perceived to have superior knowledge, experience, or judgment that others need and do not possess themselves. Referent power comes from being respected, likable, and worthy of emulating.

Recently, researchers have noted that individual power can be dichotomized into two major dimensions: position power and personal power (Yukl, 1994). Position power is based on one's formal position in an organization and is comprised of three of the five bases of power identified by French and Raven (legitimate power, reward power, and coercive power). The remaining sources of power (i.e., referent and expert) includes power derived from personal characteristics.

Other Bases of Power

Recognition of other sources of power has emerged. For instance, Yukl and Falbe (1991) have noted that information, persuasiveness, and charisma are likely bases of power. Crozier (1964) recognized task interdependence (where two or more workers must depend on one another) as a source of power. Examples include a subordinate providing a supervisor with information needed to do the job; a subordinate having the ability to use facts and logic to present a case persuasively; and an executive depending on a subordinate to get the job done correctly. In each case, the subordinate has power outside of French and Raven's five bases of power. While inclusion of these additional sources provides a more thorough and accurate representation of the types of power that exist within organizations, it is still too soon to tell whether this more comprehensive model is any better than the original. For example, while the inclusion of charisma may provide additional insights, it may prove difficult to distinguish it from referent power. Similarly, position power and legitimate power also appear to be closely related.

As such, the conceptual clarification of the original model may prove more worthwhile when investigating the power bases likely found in SMT.

Power and Influence in Organizations

As noted earlier, preferences patterns for type of influence tactics used remain fairly consistent, regardless of the agent's hierarchical position relative to the target. A similar pattern was found in the power literature; that is, the use of personal power did not vary significantly with direction of influence (Hinkin & Schriesheim, 1990). More recent research has found that, in supervisor/subordinate relationships, a leader's use of influence tactics is guided by their respective sources of power (Yukl & Falbe, 1991). Accordingly, theorists have suggested that the use of "strong arm" or coercive tactics is, for the most part, negatively related to expert and referent power (French & Raven, 1959). Specifically, Thompson (1967) found that "experts" need not use threats but, instead, use reason and logic. Similarly, it is reasonable to expect agents who have the capacity to influence SMT members run the risk of losing this power if they elect to use coercive influence tactics. A logical extension of this research is to suggest that the use of coalitions and exchange (examples of more coercive tactics shown in Table 2) might be used as a last resort when an agent only possesses personal sources of power.

In contrast, current research has consistently found that cooperative influence tactics are most commonly used when seeking to enhance personal sources of power as well as get desired outcomes (Yukl & Falbe, 1991). One explanation for the association between cooperative influence tactics and personal sources of power centers around the egalitarian (AKA consultative) nature of these influence tactics. It seems reasonable to expect more egalitarian tactics would be used when the agent has no position power and is relying on personal source of power. Research has supported this assertion. For example, Hinkin and Schriesheim (1990) found that the use of rationality was positively associated with expert and referent power. Surprisingly, no relationship was found between ingratiation and inspirational appeal (two of the more consultative tactics).

Power and Influence in SMTs

We anticipate a similar selection of influence tactics will emerge in SMTs. As noted earlier, the more consultative influence tactics have been shown to be most popular when expert and referent power are the only sources of influence. Yet, given the unique interdependent nature of SMTs, another source of power may prevail. We believe there is greater pressure in SMTs to seek mutual gain rather than playing "winner takes all." This give and take serves as the foundation for reciprocal alliances.

In Figure 1, a taxonomy of these sources of power (personal and reciprocal alliances) is presented (Level III). A proposed pairing of power bases and influence strategies is illustrated. Using the research patterns found in the literature cited above, the model suggests that more consultative influence tactics will be used when the SMT agent has personal sources of power. On the other hand, we believe more coercive tactics will be used when the SMT agent relies on reciprocal alliances to influence training decisions.

Which choice patterns SMT members will use when seeking to influence training decisions is also affected by the cohesion of the team. A discussion of the nature of team cohesion is next, followed by an analysis of its impact on SMT agent's influence choice decisions. Finally, Level IV (the impact of team cohesion) of the contingency model is addressed.

TEAM COHESION

In the SMT arena, team members work exclusively with their current team members to complete the team's duties. This level of interaction encourages the emergence of a highly cohesive team. Indeed, several researchers have argued that high levels of cohesiveness are very likely to emerge with SMTs (Barker, 1993; Goodman et al., 1988; Lawler, 1986; Manz & Neck, 1995; Manz & Sims, 1982). However, one issue of primary importance which has not been discussed is the type of cohesion that may result from this interdependence within self-managing teams.

Specifically, group cohesiveness has often been defined as the result of all the forces acting on the members to remain in the group (Festinger, 1954). Janis (1982) described cohesiveness from an interpersonal viewpoint. He argued that the more amiability and esprit de corps among the members of a group, the greater the danger that independent critical thinking is replaced by the concurrence seeking of groupthink. However, Janis's view of group cohesion is a unidimensional perspective, which may be too limiting to help us understand the relationship between influence tactics and training attendance decisions.

Previous group-processes research has indicated that cohesion is a multifaceted construct composed of task and interpersonal dimensions (Mullen & Copper, 1994; Tziner, 1982; Zaccaro & Lowe, 1988; Zaccaro & McCoy, 1988). Task-based cohesion exists when there is a shared commitment to the task of the group (Hackman, 1986). Interpersonal (social) cohesion is based on personal relationships and friendships with other members of the group (Festinger, Schachter, & Back, 1950). Research has also demonstrated that these dimensions of cohesion are independent constructs and that the two different types of cohesiveness have different effects on group norm development and productivity (Zaccaro & McCoy, 1988). This discussion suggests, therefore, that *both* the *level* of cohesion as well as the *type* of cohesion (Zaccaro & Lowe, 1988) should be considered when examining the relationship between influence tactics and training attendance decisions.

The authors who suggest that high levels of cohesiveness will emerge in SMTs do not directly differentiate between the two types of cohesiveness. Inferences can be drawn as to what form of cohesiveness might emerge in SMTs, however. For example, Manz and Sims (1982) state that there is a high level of interaction and support with other members as the team is largely responsible for managing the activities of its own group. This contention seems to point toward a task-based form of cohesiveness emerging within the team. Additionally, Lawler (1986) indicates the emergence of interpersonal cohesion within SMTs by noting that as a work team develops, membership in the team becomes very important to the workers. The group setting meets individuals' needs for social interaction and belonging. A similar interpersonal cohesion view of SMTs is apparent in Barker's (1993) longitudinal ethnography of an organization's transformation from a traditional approach to job design to a team-based design. Barker describes how new values emerged and directed team member behavior. For example, one team member talks about viewing the company as a family and her teammates as family members. Therefore, the SMT research provides evidence that both types of cohesiveness can develop within SMTs as a result of the high degree of interaction required.

Team Cohesion and Social Influence

In this section, we will explore which of Yukl & Falbe's eight influence strategies would be used when a team is experiencing either task cohesion or social cohesion. No research was found examining the relationship between types of cohesion and influence strategies. Recall that an agent's choice of influence tactic is partially determined by the anticipated reaction of the target. In a team environment, we hypothesize that an individual team member selects a tactic which corresponds with the degree of social cohesion. For example, a team member will select a cooperative tactic when their team is bound together primarily by social ties. On the other hand, a more coercive tactic will be utilized in situations of low social cohesion.

CONCLUSION

The benefits of SMT for organizations are substantial. Because SMTs have raised productivity, on average, 30 percent or more and substantially raised quality (Hackman, 1990), they appear to be the wave of the future. Given their increasing popularity and importance, further study of their dynamics, especially the antecedents and consequences of their decision making, is warranted.

We have known for some time that the same type of influence tactic will not be effective or preferred in all situations. A situational model of social influence in SMTs was presented to begin targeting preferred tactics. This model identified three determinants of a SMT agent's influence choices: the influence agent's source of power, the nature of the team's cohesion, and the content of the training request.

First, our model is one of the initial attempts to extrapolate from the traditional influence literature to the SMT context. We posited that members of SMTs would choose influence patterns similar to those found in the literature examining influence among traditional peers, subordinates, and supervisors. Further study of this assumption is imperative. We need to learn what determines effective or ineffective influence attempts in SMTs and to what extent social influence is a product of personal characteristics of the agent or other situational factors.

Next, the conceptual ideas discussed here should expand the body of knowledge involving training enrollment decisions, especially the perceived utility of two types of training. Further comparative study of these two types of training may have significant implications for practitioners. In particular, practitioners may gain insights regarding the promotion, design, and evaluation of training for SMT members. This chapter has also focused on *how* training decisions in SMTs are influenced, specifically, which influence tactics are used by SMT agents when attempting to influence training decisions. After considering our model, SMT members may acquire skill-based knowledge necessary to promote training decisions they believe are important to the success of the team.

The impact of team cohesion still needs to be explicated. Our discussion provides additional support for the call for improved assessment of team environmental factors. Still less is known about how team performance (in particular team decision making) is affected by social compared to task cohesion conditions.

We know that relationships in SMTs are complex and that group performance depends on the team's ability to coordinate its efforts and make effective decisions. Given the estimates of increasing utilization of SMTs in organizational life, the understanding of within-team processes such as training influence tactics, will hopefully offer at least a partial blueprint for enhancing employee output as we head into the twenty-first century.

REFERENCES

Albrecht, C., Jr. (1996). Career centers promote employability. *HRMagazine, 41*(8), 105-108.

Ansari, M.A., & Kapoor, A. (1987). Organizational context and upward influence tactics. *Organizational Behavior and Human Decision Processes, 40,* 39-49.

Baldwin, T.T., Magjuka, R.J., & Loher, B.T. (1991). The perils of participation: Effects of choice of training on trainee motivation and learning. *Personnel Psychology, 44,* 51-65.

Barker, J.R. (1993). Tightening the iron cage: Concertive control in self-managing teams. *Administrative Science Quarterly, 38,* 408-437.

Barry, B., & Bateman, T. (1992). Perceptions of influence in managerial dyads: The role of hierarchy, media, and tactics. *Human Relations, 45,* 555-574.

Brousseau, K.R., Driver, M.J., Eneroth, K. and Larsson, R. (1996). Career pandemonium: Realigning organizations and individuals. *Academy of Management Executive, 10,* 52-66.

Campion, M.A., Medsker, G.J., & Higgs, A.C. (1993). Relations between work group characteristics and effectiveness: Implications for designing effective work groups. *Personnel Psychology, 46,* 823-847.

Cannon-Bowers, J.A., Tannenbaum, S.I., Salas, E., & Volpe, C.E. (1995). Defining competencies and establishing team training requirements. In R.A. Guzzo, E. Salas, & Associates (Eds.), *Team effectiveness and decision making in organizations* (pp. 333-380). San Francisco: Jossey-Bass.

Cianni, M., & Wnuck, D. (1997). Individual growth and team enhancement: Moving toward a new model of career development. *Academy of Management Executive, 11,* 105-115.

Cohen, A.R., & Bradford, D.L. (1989). Influence without authority: The use of alliances, reciprocity, and exchange to accomplish work. *Organizational Dynamics, 17,* 4-17.

Cohen, S.G., & Denison, D.R. (1990). Flight attendant teams. In J.R. Hackman (Ed.), *Groups that work (and those that don't)* (pp. 382-397). San Francisco: Jossey-Bass.

Cohen, S.G., & Ledford, G.E. (1994). The effectiveness of self-managing teams: A quasi-experiment. *Human Relations, 47,* 13-43.

Cohen, S.G., Ledford, G.E., & Spreitzer, G.M. (1996). A predictive model of self-managing team effectiveness. *Human Relations, 49,* 643-676.

Corderey, J.L., Mueller, W.S., & Smith, L.M. (1991). Attitudinal and behavioral effects of autonomous group working: A longitudinal field study. *Academy of Management Journal, 34,* 464-476.

Crozier, M. (1964). *The bureaucratic phenomenon.* Chicago: University of Chicago Press.

Cummings, T.G. (1978). Self-regulation work group: A sociotechnical synthesis. *Academy Management Review, 3,* 625-634.

Cummings, T.G., & Griggs, W.H. (1977). Worker reactions to autonomous work groups: Conditions for functioning, differential effects and individual differences. *Organization and Administrative Sciences, 7,* 87-100.

Emery, F.E., & Trist, E.L. (1969). Socio-technical systems. In F.E. Emery (Ed.), *Systems thinking* (pp. 281-296). London: Penguin Books.

Erez, M., Rim, Y., & Keider, I. (1986). The two sides of the tactics of influence: Agent vs. target. *Journal of Occupational Psychology, 59,* 25-39.

Festinger, L. 1954. A theory of social comparison processes. *Human Relations, 7,* 117-140.

Festinger, L., Schachter, S., & Back, K. (1950). *Social pressures in informal groups: A study of human factors in housing.* New York: Harper Row.

Fiedler, F.E. (1964). A contingency model of leadership effectiveness. In L. Berkowitz (Ed.), *Advances in experimental social psychology.* New York: Academic Press.

French, J.R., & Raven, B.H. (1959). The bases of social power. In D. Cartwright (Ed.), *Studies of social power* (pp. 150-167). Ann Arbor, MI: Institute for Social Research.

Goldstein, I.L., & Gilliam, P. 1990. Training system issues in the year 2000. *American Psychologist, 45,* 134-143.

Goodman, P.S., Devadas, R., & Hughson, T.L.G. (1988). Groups and productivity: An analysis of self-managing teams. In J.P. Campbell (Ed.), *Productivity in organizations* (pp. 295-327). San Francisco: Jossey-Bass.

Hackman, J.R. (1986). The psychology of self-management in organizations. In M.S. Pollack & R.O. Perlogg (Eds.), *Psychology and work: Productivity change and employment* (pp. 85-136). Washington.

Hackman, J.R. (1987). The design of work teams. In J.W. Lorsch (Ed.), *Handbook of Organizational Behavior* (pp. 315-342). Englewood Cliffs, NJ: Prentice Hall.

Hackman, J.R. (1990). *Groups that Work (and Those that Don't).* San Francisco: Jossey-Bass.

Herbst, P.G. (1962). *Autonomous group functioning and exploration in behavior theory and measurement*. London: Tavistock.

Hinkin, T.R., & Schriesheim, C.A. (1990). Relationships between subordinate perceptions of supervisor influence tactics and attributed bases of supervisory power. *Human Relations, 43,* 221-237.

House, R.J. (1971). A path-goal theory of leader effectiveness. *Administrative Science Quarterly, 16,* 321-339.

Janis, I.L. (1982). *Groupthink.* Boston: Houghton Mifflin.

Katzenbach, J.R., & Smith, D.K. (1993). *The wisdom of teams: Creating the high performance organization.* Boston: Harvard School Press.

Kipnis, D., & Schmidt, S.M. (1988). Upward-influence styles: Relationship with performance evaluations, salary, and stress. *Administrative Science Quarterly, 33,* 528-542.

Kipnis, D., Schmidt, S.M., & Wilkinson, I. (1980). Intra-organizational influence tactics: Explorations in getting one's way. *Journal of Applied Psychology, 65,* 440-452.

Koonce, R. (1996). Ensuring your employability. *Training & Development, 50*(7), 14.

Lawler, E.E. (1986). *High involvement management.* Jossey-Bass: San Francisco.

Lawler, E.E. (1993). Creating the high-involvement organization. In J.R. Galbraith, E.E. Lawler, & Associates (Eds.), *Organizing for the future: The new logic for managing complex organizations* (pp. 172-194). San Francisco: Jossey-Bass Publishers.

Manz, C.C. (1992). *Mastering self-leadership*. Englewood Cliffs, NJ: Prentice Hall.

Manz, C.C., & Angle, H. (1986). Can group self-management mean a loss of personal control: Triangulating on a paradox. *Group and Organization Studies, 11,* 309-334.

Manz, C.C., Keating, D., & Donnellon, A. (1990). Preparing for an organizational change to employee self-management: The managerial transition. *Organizational Dynamics, 19,* 15-26.

Manz, C.C., & Neck, C.P. (1995). Teamthink: Beyond the groupthink syndrome in self-managing teams. *Journal of Managerial Psychology, 10,* 7-15.

Manz, C.C., Neck, C.P., Mancuso, J., & Manz, K. (1997). *For team members only: Making your workplace team productive and hassle-free*. New York: Amacom.

Manz, C.C., & Newstrom, J. (1990). Self-managing teams in a paper mill: Success factors, problems, and lessons learned. *International Human Resource Management Review, 1,* 43-60.

Manz, C.C., & Sims, H.P., Jr. (1982). The potential for groupthink in autonomous work groups. *Human Relations, 35,* 773-784.

Manz, C.C., & Sims, H.P., Jr. (1987). Leading workers to lead themselves: The external leadership of self-managed work teams. *Administrative Science Quarterly, 32,* 106- 128.

Manz, C.C., & Sims, H.P., Jr. (1993). *Business without bosses: How self-managing teams are building high performance companies*. New York: Wiley.

Manz, C.C., & Stewart, G.L. (1997). Attaining flexible stability by integrating total quality management and socio-technical systems theory. *Organizational Science, 8,* 59-70.

Mintzberg, H. (1983). *Power in and around organizations*. Englewood Cliffs, NJ: Prentice Hall.

Mirvis, P.H., & Hall, D.T. (1994). Psychological success and the boundaryless career. *Journal of Organizational Behavior, 15,* 365-380.

Mohrman, S.A., & Mohrman, A.M., Jr. (1993). Organizational change and learning. In J.R. Galbraith, E.E. Lawler, & Associates (Eds.), *Organizing for the future: The new logic for managing complex organizations* (pp. 87-108). San Francisco: Jossey-Bass Publishers.

Moorhead, G. Neck, C.P., & West, M. (1996). The tendency toward defective decision making within self-managing teams. Paper presented at the Academy of Management Meetings, Cincinnati, Ohio.

Mullen, B., & Copper, C. (1994). The relation between group cohesiveness and performance: An integration, *Psychological Bulletin, 115,* 210-227.

Orsburn, J., Moran, L., Musselwhite, E., Zenger, J., & Perrin, C. (1990). *Self-directed work teams: The new American challenge*. Homewood, IL: Business One Irwin.

Pfeffer, J. (1981). *Power in organizations*. Marshfield, MA: Pitman.

Pfeffer, J. (1992). *Managing with power: Politics and influence in organizations.* Boston, MA: Harvard Business School Press.

Poza, E.J., & Markus, M.L. (1980, Winter). Success story: The team approach to work restructuring. *Organizational Dynamics, 8,* 3-27.

Salas, E., Dickinson, T.L., Converse, S.A., & Tannenbaum, S.I. (1992). Toward an understanding of team performance and training. In: R.W. Swezey & E. Salas (Eds.), *Teams: Their training and performance* (pp. 3-29). Ablex Publishing: Norwood, NJ.

Saporito, B. (1986). The revolt against working smarter. *Fortune, 114*(2), 58-65.

Schmidt, S.M., & Kipnis, D. (1984). Manager's pursuit of individual and organizational goals. *Human Relations, 37,* 781-794.

Shaw, B. (1990). Mental health treatment teams. In J.R. Hackman (Ed.), *Groups that work (and those that don't)* (pp. 330-348). San Francisco: Jossey-Bass.

Sims, H.P., Jr., Manz, C.C., & Bateman, B. (1993). The early implementation phase: Getting teams started in the office. In C.C. Manz & H.P. Sims, Jr., (Eds.), *Business without bosses: How self-managing teams are building high performing companies* (pp. 85-114). New York: Wiley.

Susman, G.I. (1976). *Autonomy at work: A socio-technical analysis of participative management.* New York: Praeger.

Tannenbaum, S.I., Beard, R.L., & Salas, E. (1992). Team building and its influence on team effectiveness: An examination of conceptual and empirical developments. In K. Kelley (Ed.), *Issues, theory, and research in industrial/organizational psychology* (pp. 117-153). Amsterdam: Elsevier Science Publishers.

Thompson, J.D. (1967). *Organizations in Action.* New York: McGraw Hill.

Trist, E., Susman, G.I., & Brown, G.R. (1977). An experiment in autonomous working in an American underground coal mine. *Human Relations, 30,* 201-236.

Tziner, A. (1982). Differential effects of group cohesiveness types: A clarifying overview. *Social Behavior and Personality, 10,* 227-239.

U. S. Department of Labor (1993). *High performance work practices and firm performance.* Washington, DC: Office of the American Workplace.

Verespej, M.A. (1990). Self-directed work teams yield long-term benefits. *Journal of Business Strategy, 11*(6), 9-12.

Walton, R.E. (1977). Work innovations at Topeka: After six years. *Journal of Applied Behavioral Science, 13,* 422-433.

Walton, R.E. (1985). From control to commitment in the workplace. *Harvard Business Review, 63*(2), 76-84.

Yukl, G. (1989). *Leadership in organizations* (2nd ed.) Englewood Cliffs, NJ: Prentice Hall.

Yukl, G. (1994). *Leadership in organizations* (3rd ed.) Englewood Cliffs, NJ: Prentice Hall.

Yukl, G., & Falbe, C.M. (1990). Influence tactics and objectives in upward, downward, and lateral influence attempts. *Journal of Applied Psychology, 75,* 132-140.

Yukl, G., & Falbe, C.M. (1991). Importance of different power sources in downward and lateral relations. *Journal of Applied Psychology, 76,* 416-423.

Yukl, G., Falbe, C.M., & Youn, J.Y. (1993). Patterns of influence behavior for managers. *Group & Organization Management, 18,* 5-28.

Yukl, G., & Tracey, J.B. (1992). Consequences of influence tactics used in subordinates, peers, and the boss. *Journal of Applied Psychology, 77,* 525-535.

Zaccaro, S.J., & Lowe, C.A. (1988). Cohesiveness and performance on an additive task: Evidence for multidimensionality. *Journal of Social Psychology, 128,* 547-558.

Zaccaro, S.J., & McCoy, C. (1988). The effects of task and interpersonal cohesiveness on performance of a disjunctive task. *Journal of Applied Social Psychology, 18,* 837-851.

Zenger, J.H., Musselwhite, E., Hurson, K., & Perrin, C. (1994). *Leading teams: Mastering the new role.* Homewood, IL: Business One.

EVALUATING THE CAPABILITY FORMATION POTENTIAL OF INTEGRATED PROJECT TEAMS

Paul F. Skilton and Vicki Smith-Daniels

ABSTRACT

This chapter develops theory about the relationship between specific types of learning by participants in integrated project teams and the formation of new strategic capabilities. Drawing on an exploratory study of new product development teams and theory from several areas of research we define a T-shaped set of skills as necessary to forming new strategic capability. Opportunities for learning T-shaped skills are primarily determined by whether the organization of work in integrated project teams reduces, emphasizes, or enhances the task interdependence encountered by participants. We develop a model of the relationship between the frequency of opportunities for desirable types of participant learning, task interdependence, and the characteristics of the integrated project team. Finally we use this model to develop a qualitative framework for evaluating and managing the capability development potential of integrated project teams. We close with suggestions for future research.

Advances in Interdisciplinary Studies of Work Teams, Volume 6, pages 91-115.

ISBN: 0-7623-0655-6

INTRODUCTION

Employees benefit from participation in integrated project teams (IPTs), because it exposes them to opportunities to learn, grow, and change (Harris & DeSimone, 1994; Katzenbach & Smith, 1994; Sims & Manz, 1994; Tjosvold & Tjosvold, 1995). An IPT is a multidisciplinary, relatively autonomous, project-oriented work team (Department of Defense, 1996). Specific types of employee learning as a result of participation in new product development IPTs are widely regarded as critical to the formation of strategic product development capabilities (Iansiti, 1993; Klein & Maurer, 1995; Leonard-Barton, 1995; Nonaka, 1994; Prahalad & Hamel, 1990; Wheelwright & Clark, 1992). In spite of high levels of agreement about these two points, theory development about how opportunities for specific types of participant learning in IPTs can be created and managed to facilitate capability formation has been lacking. By filling this gap we contribute to an understanding of how micro-level participant learning in the IPT context leads to the macro-level formation of a sustained capability.

More critical than the gap in theory is the gap in practice. New product development managers have tended to lose sight of both participant learning and the formation of strategic capabilities in the urgent business of pushing the right new product out to market on time. Creating opportunities for participant learning to support strategic capability formation is often a distant, secondary goal compared to getting the primary jobs of the IPT done. This is a dangerous omission when new product development is a vital strategic capability. We contribute to closing this gap by developing a qualitative framework for evaluating and managing integrated project teams to support strategic capability formation. This framework takes into account both the necessity to achieve desirable primary outcomes from projects, and the necessity to promote participant learning.

Our theory development in this chapter is grounded in observations made while collecting exploratory data for a study of new product development IPTs in large high tech companies. We used the results of our fieldwork to guide our choice of theories and constructs to build up a syncretic theory of the relationship between capability development and participant learning. The eight new product and process development IPTs we studied were selected to expose us to successful teams who represented the leading edge of integrated process and product development in their respective companies. The parent organizations were all participants in the aerospace industry; projects ranged from whole aircraft to small components and test equipment.

Figure 1 presents a road map of our theory. Strategic capability formation is achieved through two sets of integrated project team outcomes. Previous work on strategic capability development has been dominated by an emphasis on what we call *primary outcomes*: speed of development, cost reduction, improvements in quality, reliability and performance, integration of new technology, and so on. In general, achieving these outcomes is a result of achieving fit between the level of

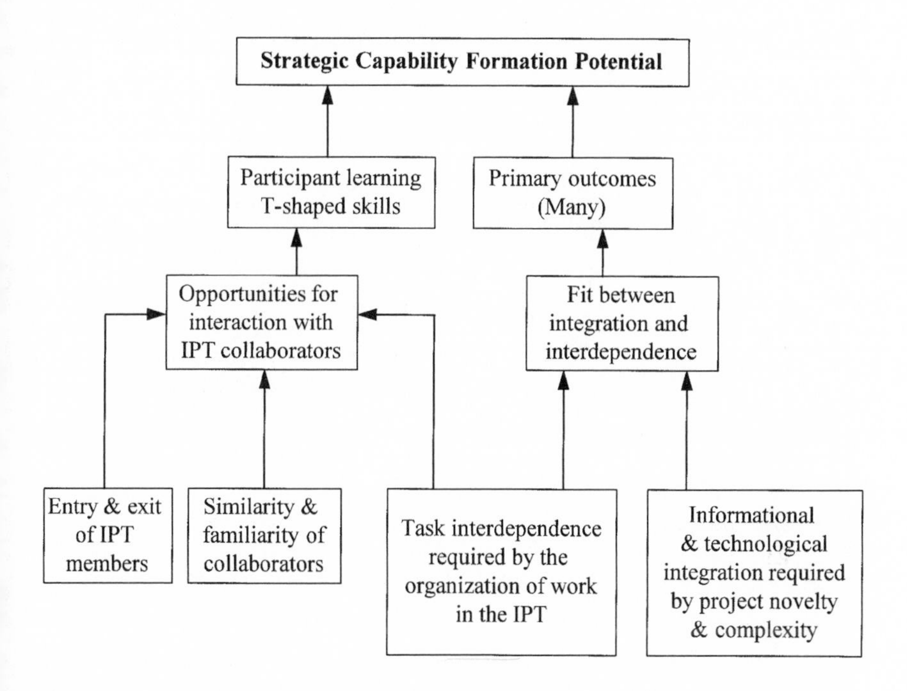

Figure 1. The Antecedents of Strategic Capability Formation through IPT

integration required by the novelty and complexity of a project, and the level of task interdependence established by the organization of work in the team. If high levels of integration are necessary, it is helpful to organize work to achieve high task interdependence between participants.

Our fieldwork leads us to believe that the organization of work in IPTs is key to participant learning as well as essential to primary outcomes. Not only should the organization of work produce the level of integration required by the project, it should also contribute significantly to the production of learning opportunities for IPT participants. The level of task interdependence built into the organization of work in an IPT controls the level of interaction between participants, and hence determines whether or not specific types of learning opportunities will occur. Where work is organized to enhance task interdependence, the level of interaction between participants and the occurrence of opportunities for learning T-shaped skills will be increased. Based on our fieldwork, we feel safe in assuming that participants will exploit opportunities to learn as much as possible.

As shown in Figure 1, the second basis for strategic capability formation is the specific types of participant learning that make up the T-shaped skill set. Starting from Iansiti's (1993) idea of T-shaped skill sets, we make a further contribution

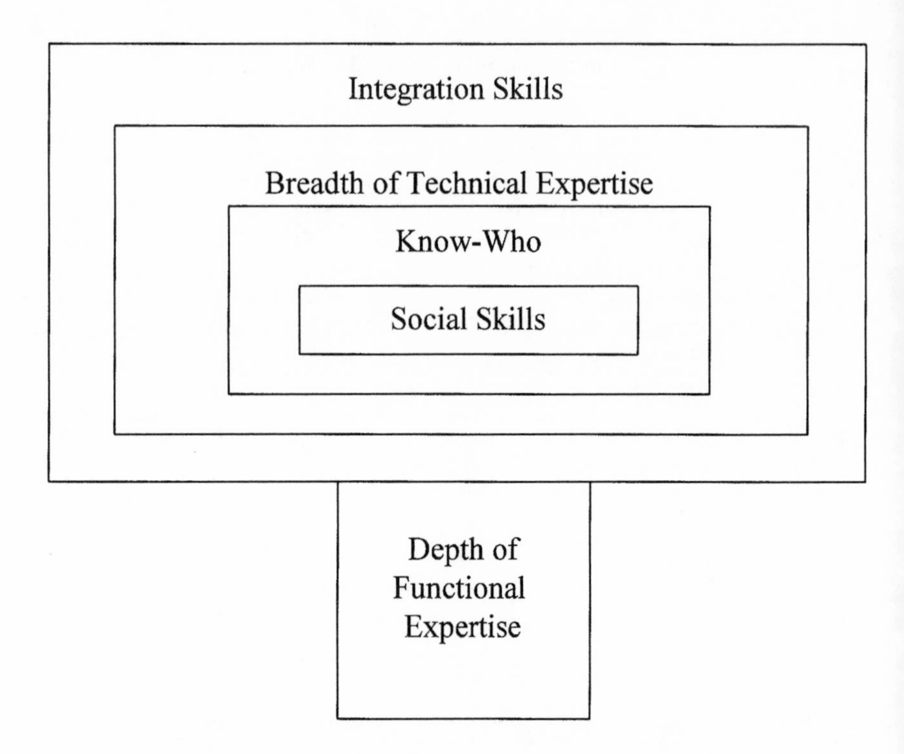

Figure 2. The Components of T-Shaped Skills

by refining the skill set that contributes to capability development. We propose to add two skills to the T-shaped skill set: "know-who" and inclusive, "voice" oriented social skills. These have been primarily studied in the laboratory, but our exploratory fieldwork leads us to believe that they play a critical role in capability formation. Along with integrative skills and breadth and depth of technical knowledge, they are skills learned in and contributing to collective integration of multiple perspectives (Boland & Tenkasi, 1995), technologies and areas of expertise (see Figure 2).

In addition to the organization of work, the occurrence of opportunities for participant learning in IPTs is also influenced by the movement of participants in and out of the team, and the experience levels of participants. In the teams we studied, the movement between the functional organization and teams was an essential source of specialist knowledge for the teams. When we consider the evaluation and management of IPTs, we need to keep in mind the novelty and complexity of the project and its influence both on the organization of work and the achievement of primary outcomes.

The object of this chapter is to develop theory about the relationship between specific types of learning by participants in integrated project teams and the *for-*

mation of strategic capabilities. Not all organizations are in the developmental stage of capability formation. There are companies who have achieved very mature capabilities. This chapter is geared toward companies that are less mature, that are still developing capabilities. We think that companies with mature capabilities could apply the theory and evaluative framework we present here when they engage in the process of renewing and reframing capabilities, but our primary intention is to provide a tool for building capabilities where they don't currently exist. We accomplish this by drawing on the experience of teams we have studied in the field to find connections between theories from multiple disciplines. The result is a syncretic theory about the connections between the organization of team work, specific types of learning through participation in IPTs, and the formation of organizational capabilities. By understanding the types of learning that matter and how they relate to the organization of team work, managers can evaluate and manage IPTs to support strategic capability formation.

STRATEGIC CAPABILITY FORMATION

The resource-based view of business strategy (Barney, 1991; Collis & Montgomery, 1995; Conner, 1991; Peteraf, 1993) is dominated by the idea that some organizational capabilities are the source of competitive advantage. A *capability* is "the capacity for a set of resources to integratively perform a task or activity" (Hitt, Ireland, & Hoskisson, 1997, p. 21). A *core competence* (Prahalad & Hamel, 1990) is a capability that creates sustained competitive advantage, usually through the development and introduction of core products and product lines. The formation of core competencies involves learning how to coordinate diverse production skills and integrate multiple streams of technologies in order to respond quickly and efficiently to the changing product needs of a market (Prahalad & Hamel, 1990; see also Leonard-Barton, 1995; Hitt, Hoskisson, & Nixon, 1993). Therefore, there are multiple objectives associated with any capability: speed, cost leadership, quality, and product performance to name a few.

Like capabilities, core competencies integrate different resources and involve multiple objectives. Unlike common capabilities, core competencies are rare, valuable, and hard to imitate (Barney, 1991; Collis & Montgomery, 1995). They are most commonly rare and hard to imitate because they "exist in a complex web of social interactions, and may even depend on particular individuals" (Collis & Montgomery, 1995, p. 163). The test of value depends on performance in a competitive marketplace. Since evaluating competitive performance is beyond the scope of this chapter, we focus on "strategic capabilities" rather than core competencies. We define *strategic capabilities* as capabilities that are part of the "price of admission" to the competitive arena. An organization must have such capabilities in order to compete. New product development is an example of a strategic capability necessary to compete in many markets.

Strategic capabilities, like core competencies, form as organizations learn to accomplish inter-functional integration (Hitt et al., 1993; Prahalad & Hamel, 1990). Glynn, Lant, and Miliken (1994) argue that organizational learning requires a series of cross-level processes. In brief, they propose that organizational learning leads to organizational action, which forms an organizational context for individual learning. Individual learning leads back to organizational learning through processes of diffusion and institutionalization. In terms of this chapter, organizations learn about new ways to organize work (e.g., IPTs), and take organizational action to apply them. Novel organizational forms then provide a context for individual learning, which feeds back to organizational learning by modifying the future use of the form. As organizations repeatedly move through this cycle, the organizational capability matures. There are other organizational designs that facilitate individual learning, such as the Japanese practice of designing career paths to include assignment to multiple functional areas in succession (Ward, Liker, Cristiano, & Sobek, 1995), but IPTs have the advantage of being a context that is geared toward rapid inter-functional integration where little previously exists.

The question of how to form strategic capabilities through the use of IPTs is therefore the question of how to retain the lessons learned by participants in IPTs and diffuse them to the rest of the organization, especially to participants in future IPTs. The teams literature (e.g., Katzenbach & Smith, 1994) and the theoretical strategic product-development literature (Bowen, Clark, Holloway, & Wheelwright, 1994a; Hitt et al., 1993; Wheelwright & Clark, 1992) assume that IPTs are contexts in which participants always learn, that lessons learned will be desirable, and that they will be diffused to the rest of the organizations by the reassignment of participants to follow-on projects. The empirical literature on strategic product development suggests that this has not usually been the case (Bowen, Clark, Holloway, & Wheelwright, 1994b; Donnellon, 1993; Dougherty & Hardy, 1996). Less mature companies don't implement devices to evaluate the quality of learning (like post-project reviews or project audits), or evaluate and reassign participants to projects in a coherent way (Klein & Maurer, 1995). In order for long-term benefits from IPTs to occur, they must be managed not only to produce desirable project outcomes, but also to ensure that participants learn, retain, and pass on useful skills and knowledge.

Integrated Project Teams and Multiple Primary Outcomes

For the organizations we studied, developing new products with IPTs appears to reduce development time, to lower development and total costs, to increase innovation, quality, reliability, market success, and customer satisfaction (Wheelwright & Clark, 1992). The popularity of IPTs in the companies we have been involved with has been reinforced by the interest of customers and important stakeholders, who benefit directly from these primary outcomes (Department of

Defense, 1996). Interest from many sources helps multiply the number of demands placed on IPTs and IPT participants. The IPTs we studied were selected primarily because of their success in producing multiple desirable outcomes.

IPTs are not always successful in producing desirable primary outcomes. Formation of strategic capabilities requires an organization to move beyond occasional success to be able to reproduce desirable outcomes in a sustained way. While we concentrate on participant learning as the major process that leads to sustainable capability it is essential to point out that IPT success in primary outcomes is an indispensable part of capability development. It is not enough for participants to learn what *not* to do in an IPT—they must learn to actually coordinate diverse production skills and integrate multiple streams of technologies. If using IPTs to develop products and processes (or any other method of organizing work) does not produce desirable results, it is likely to be abandoned. Participants may learn desirable skills in failed IPTs, but those skills will not be redeployed unless there is some belief in the organization that using IPTs will pay off. Furthermore, IPTs can produce successful outcomes without providing much opportunity for participant learning. A dominant leader with strong integrative skills, for instance, can organize work so that participants are isolated and still generate desirable outcomes. Thus primary outcome success and participant learning in IPTs are loosely coupled—participant learning does not necessarily generate success, although success leads to more use of IPTs, and hence to more learning opportunities.

Finally, not all desirable primary outcomes are created equal. What constitutes a good primary result for an organization in the early stages of capability formation is not the same as a good primary result when a capability is mature. For example, very large proportional reductions in cycle time are possible when a capability is forming, but the same absolute cycle time may be unimpressive if applied to a more mature organization.

Participant Learning: T-Shaped Skills

The formation of strategic capabilities requires individuals to learn how to coordinate diverse productions skills and integrate multiple streams of technologies. To accomplish this participants need to acquire depth of expertise in a specialized skill set (the skills of the functional "silo") *and* the lateral skills necessary to adapt and integrate their own expertise with that of diverse participants (Barley & Orr, 1997; Klein, 1994; Klein & Maurer, 1995; Hitt et al., 1993; Leonard-Barton, 1995; Nonaka, 1994). Iansiti (1993) and Leonard-Barton (1995) called this aggregate skill set "T-shaped skills," where for each individual, the stem of the "T" is deep skill in a functional area and the crossbar is composed of the set of skills necessary to adapt to and integrate one's own skill set with others. The specific types of learning involved in the acquisition of T-shaped skills are noted in Figure 2 and addressed in detail below. The opportunity to acquire T-shaped skills is the critical opportunity created by participation in IPTs.

IPT participants are exposed to the opportunity to develop depth of expertise in functional skills (Klein, 1994) because the tasks of IPTs often require new applications of functional skills. Each new application or problem encountered expands the body of expertise. Depth of expertise can also be acquired as a participant learns how his or her functional area interfaces with new disciplines and technologies. It is, of course, not necessary to participate in IPTs to be exposed to new applications of functional skills or develop depth of expertise. In our fieldwork informants often reported that while they felt they *could* develop functional expertise in conventional departments, they were more likely to do so in situations requiring intensive interaction and collaboration. This is in contrast to the concerns of some researchers (Leonard-Barton, 1995) that removing skilled individuals from a conventional specialist department diminishes the depth of knowledge available to the organization. Several team members reported that their home department managers had a parallel concern, to the extent that the team members hid the scope of their involvement in different aspects of the project from those managers.

There are four components of crossbar skills. These four components are closely related, but sufficiently different to warrant separate treatment. As shown in Figure 2, the four components can be conceived of as nested, with the interior skills being amplified and enhanced by the presence of outer skills. The four components include the following.

Integration Skills

Klein (1994) argues that integration expertise "does not necessarily require in-depth operational expertise or analytic knowledge concerning the tasks (or skills) to be integrated" (p. 149). It does require breadth of technical knowledge to be able to recognize problems. Beyond breadth of knowledge, integration skills require expertise in architectural (Henderson, 1996) or systems knowledge of the project and its requirements. There is a trend in contemporary organizations to use expert systems, knowledge bases, project management software, and other coordinating mechanisms to support the integration of diverse production skills and multiple technologies. The function of these coordinating mechanisms is to force IPT participants to engage in integrative activities early and learn integration skills. Integration skills in the aerospace industry have traditionally been the domain of the systems integration or configuration management department; an important trend we observed was the incorporation of integration skills into the IPTs we studied. Members of the integrative departments often served as sources of knowledge for other team members about how to integrate different functions and project elements. Learning and applying integration skills requires less intensive contact than do other crossbar skills.

Breadth of Technical Knowledge

Leonard-Barton (1995) notes that some individuals have depth of expertise in multiple disciplines as well as crossbar skills. These rare individuals are said to have "A-shaped skills" because they have expertise in more than one area. These areas form the legs of the "A," which are joined at the top by the individual's ability to translate between the areas without explicit integration skills. The crossbar of the "A" is made up of various integration skills as defined above.

More commonly participants acquire limited knowledge of the disciplines of the people they interact with and depend on. In our exploratory fieldwork we often encountered multiskilled individuals such as those described by Klein (1994) who "developed sufficient knowledge of each other's function to better integrate their skills" (p. 151). Such multiskilled participants are not necessarily multi-expert, but they know enough to be able to understand each other, and recognize ideas that fit across disciplines. Nonaka (1994) argues for the importance of multiskill development as facilitating interaction and knowledge creation. Northcraft, Polzer, Neale, and Kramer (1996) argue that redundancy of skills is required for cross-fertilization in multidisciplinary teams because it permits participants to adopt the frames of reference of their fellows, and thus engage in what Boland and Tenkasi (1995) call perspective taking. This type of learning is greatly facilitated by incidents of close contact between different participants in IPTs. Participants in the teams we studied reported that site visits and involvement in the manufacturing process, where they could experience firsthand the problems facing collaborators, were particularly useful for this kind of learning.

Know-Who Knowledge

Integration skills and breadth of expertise imply that participants "know who to call when they need an expert" (Klein, 1994, p. 151). Cohen and Levinthal make this point in their discussion of absorptive capacity: "Critical knowledge…includes awareness of where useful complementary expertise resides within and outside the organization. This sort of knowledge can be knowledge of who knows what, who can help with what problem or who can exploit new information" (1990, p. 133). Moreland, Argote, and Krishnan (1996) argue that transactive memory in groups (Wegner, 1987) facilitates "tacit coordination" and group productivity. There is evidence supporting the idea that "knowing who knows what" facilitates group performance from both laboratory (Moreland et al., 1996; Rulke & Rau, 1997; Stasser, Stewart, & Wittenbaum, 1995) and field (Goodman & Leyden, 1991). Participation in IPTs facilitates the acquisition of know-who knowledge, because it exposes participants to the organizational knowledge of people from different parts of the organization. Interaction does not need to be intense, but know-who may improve with repeated interactions, as each participant develops a better representation of what the other's areas of expertise are.

Social Skills

Beyond knowing who knows what and being able to integrate various disciplines, partly because of breadth of technical knowledge, social skills are an important component of crossbar skills. These include an understanding of group dynamics, conflict resolution, problem solving, decision making, and communication skills (Jackson, May, & Whitney, 1995; Klein, 1994; Lawler, 1992; Leonard-Barton, 1995; Orsburn, Moran, Musselwhite, & Zenger, 1990). In addition to social skills that have previously been recognized in the teams literature, we believe that a critical social skill is the ability to include and value the points of view of participants whose participation might otherwise not be seen as legitimate (Lave & Wenger, 1991). In the procedural justice literature, this concept is known as "voice" (Shapiro, 1993). In IPTs the skill of ensuring voice for all participants is important, not only because it helps create commitment and perceptions of justice, but also because if participants are mute (or muted) their knowledge and skills are only incorporated into the project with difficulty. Voice thus relates to Leonard-Barton's concept of core rigidity (1992), to Dougherty and her associates' research into the illegitimacy of innovations (Dougherty, 1992, 1996; Dougherty & Hardy, 1996; Dougherty & Heller, 1994), and Cohen and Levinthal's (1990) discussions of the "not invented here" phenomenon. Clearly before someone can legitimize and value an excluded point of view, he or she has to know that it exists, who holds it, and how it fits. Acquiring voice-enhancing social skills requires close contact and considerable effort from diverse participants.

Dimensions of Participation in IPTs

IPTs are by definition multidisciplinary, and hence made up of participants with diverse knowledge, skills, and abilities (KSAs) (Jackson et al., 1995; Northcraft et al., 1996). Most of the IPTs we studied included participants from all of the internal stakeholder constituencies that had an interest over the total project life cycle, but few included customers and none directly included suppliers. Participants were primarily engineers, although there was a significant proportion of other knowledge workers (Barley & Orr, 1997; Mohrman, Mohrman, & Cohen, 1995; Purser & Montuori, 1995) including skilled craftspeople and specialists from business disciplines. The size of the teams varied, although they were on the large side (15 to 200 members).

The importance of participation in IPTs for the development of individual employees is widely recognized (Cummings, 1977; Lawler, 1992; Manz & Sims, 1993) but it is seldom the primary consideration in selecting IPT members

(Leonard-Barton, 1995). Team leaders frequently select individuals they have worked with successfully on previous projects, or who have good reputations from previous projects. Team leaders seldom select all of the members of a team, however; department managers often choose individuals to represent a department or function on the basis of availability rather than special expertise or experience. The result of this pattern of selection is that membership in IPTs is often stratified between more experienced members, who are familiar with IPT practices (and each other), and novices. Repeated experience in IPTs usually means that a participant has already developed technical depth and social and integrative skills. Carrying crossbar skills forward means that the experienced participant has less to learn, and thus benefit from different kinds of interaction than novices do.

A balanced mix of experienced and novice members improves the capability formation potential of the IPT by providing novices with an opportunity to learn from the experience of veterans. Novices have an opportunity to acquire integration, know-who, and social skills more rapidly when in contact with experienced IPT members. Experienced IPT members should benefit from contact with novices. In the teams we studied, this often took the form of improving the depth or breadth of functional skills, as novices often possessed more recently acquired, leading-edge technical knowledge. Several teams reported that when more experienced participants recognized problems, less experienced participants often brought forward the eventual solutions.

This benefit was undermined in some of the teams we observed by the practice of beginning an IPT with a core group of experienced participants, and adding members, usually novices, later in the project life cycle. In some cases, experienced participants left the team as the project life cycle progressed. As a result, the composition of the "community of interaction" (Nonaka, 1994) in these teams was not stable over time. This has adverse consequences for the intensity of interaction of the IPT. In the teams we studied, novice participants who joined the team later in the cycle often felt confused and isolated because they had little contact with senior team members. These novices are often put to work with little understanding of the "why" behind the work they are asked to do. The senior members often recognized this, but were unable to interact with novices because of schedule demands, or because they had been reassigned. The senior members typically expressed regret and a sense of loss about this missed opportunity. In general, therefore, participants who contribute to the whole project life cycle have much better opportunities for learning than those who come and go.

Participant Learning, T-Shaped Skills, and Capability Development

Personal contact between participants is important for novices, because individual learning (particularly of crossbar skills) is not merely a disembodied, impersonal activity (Blackler, 1995; Boland & Tenkasi, 1995; Lave & Wenger, 1991; McDermott, 1995; Nonaka, 1994). Rather, individual learning is a social process.

Lave and Wenger (1991) make the distinction between "instruction" and "participation" as modes of thinking about learning processes. Instruction is associated with training as opposed to learning through participation. Employees can be instructed in specialized knowledge and in methods of collaboration, but neither the employee nor the organization can benefit from such training until the individual has a chance to use such skills by participating in collective work.

In order to understand how participation in IPTs creates opportunities for learning T-shaped skills, it is useful to distinguish between "communities of interaction" (Nonaka, 1994) and "communities of practice" (Brown & Duguid, 1991; Orr, 1990) or "communities of knowing" (Boland & Tenkasi, 1995). Individuals become members of communities through participation. When the participant is a novice, participation is geared toward learning the skills, values, beliefs, and perspectives that define the community. The novice takes part in the work, but is not expected to "know the ropes" to the same extent that experienced community members do.

Communities of practice or knowing are made up of participants from a single area, occupation, or profession: for example, doctors, pilots, engineers, or copier service technicians. Participants have varying levels of expertise in a single set of "signature skills" (Leonard-Barton, 1995). In communities of practice there are always novices and masters. Novices in communities of practice "learn the ropes" of the profession or occupation and acquire the "thought world" characteristic of the community through interaction with those who have mastered the signature skills. In Boland and Tenaksi's terms (1995), communities of practice are "perspective making"; they create a single, shared way of looking at the world. Participation in a community of practice as we have defined it is more likely to lead to an increase in the depth of functional expertise than an increase in opportunities for the other kinds of learning required for capability formation. Participants do not require integrative skills to function in a community where everyone is more or less the same, and there is little available in the way of breadth of technical knowledge. Knowledge of who knows what is likewise of diminished importance. Some social skills will be acquired through participation in a community of practice, but not the key skill of being able to value and include alternative or apparently illegitimate points of view.

IPTs have the *potential* to be communities of interaction. Communities of interaction are made up of participants who bring different functional and disciplinary skills, and different "thought worlds" (Dougherty, 1992) to the community. In communities of practice there are always novices, but it may be that no participant has mastery of the necessary skills. Novices in communities of interaction sometimes are forced to learn from and with each other. Participants develop common experience based on integrating and adapting to the differences they bring to the community. Participants are constrained and informed by each other, but they can't adopt each other's functional and disciplinary practices, because they work in different areas. Legitimate participation in a community of interaction requires what Boland and Tenkasi (1995) call "perspective taking"—taking the perspective of another into account. Full participation thus forces what Leonard-Barton (1995) calls "creative abrasion" within participants and leads them to learn about the systemic impact of different skills and preferences on the community. This process is far more effective if different perspectives and preferences can be "de-personalized," made legitimate and given voice (Dougherty & Hardy, 1996; Leonard-Barton, 1992, 1995; Shapiro, 1993).

Because IPTs have the potential to be communities of interaction, they have the *potential* to promote all of the types of learning that contribute to capability formation. Opportunities for acquiring depth and breadth of technical expertise, integration skills, know-who knowledge, and social skills are more likely to occur when participants are diverse and interact intensely. Putting together a team of people with different skills and a common task does not lead to a community of interaction. An interactive, collaborative, diverse IPT creates more intense opportunities for learning than an isolating, homogenous one does. Both types can produce desirable primary outcomes, but their impact on capability formation is very different.

Integration Requirements and Organization of Work in IPTs

Achieving the primary outcomes of IPTs depends on the novelty and/or complexity of the project and the way that work within the IPT is organized. The novelty and complexity of the project determine the technological and informational integration requirements of the work (Tushman & O'Reilly, 1997; Wheelwright & Clark, 1992). The way that the work is organized determines whether those integration requirements will be efficiently met. Thus fit between the integration requirements of a project and the organization of the project's work is critical to achieving primary outcomes.

A project that invokes many practices, technologies, and objectives is more complex than one that involves only a single technology or objective. Novelty has primarily been considered in the past as a characteristic of technologies and team practices, but Tatikonda (1995) argued that many of the objectives associated with new product development IPTs were also novel. Derivative products, familiar practices, technologies or objectives require little integration, and are usually handled by routine; novel ones require more integration. Novelty and complexity often interact. The level of complexity associated with a project may be novel, and the novelty of a component may be considered to increase the complexity of a project by rendering the relationships between project components uncertain (Henderson, 1996; Hitt et al., 1993; Iansiti, 1993). For example, when a novel objective like rapid development time is injected into an IPT, more traditional objectives like superior product performance are displaced until the relationship and priority of all objectives are reinterpreted.

The tasks of an IPT can be organized in a variety of ways. Work can be designed to be highly interdependent, requiring the cooperation and collaboration of several participants to accomplish it. Work can also be designed to be highly independent, so that individuals work alone, without much input from each other. This characteristic of work design is task interdependence (Thompson, 1967; Tushman & O'Reilly, 1997; Wageman, 1995). In Thompson's terms, work can be organized so that interdependence is pooled, sequential or reciprocal. These types of interdependence are described in Table 1. Thompson describes these types as making up a Guttman-type cumulative scale, so that if reciprocal interdependence is present, sequential and pooled are also present. For our purposes, increasing levels of task interdependence also imply a need for increased intensity of interactions between participants. When work is organized to enact pooled interdependence, producing primary outcomes does not require participants to interact. When work is organized to enact reciprocal interdependence, producing primary outcomes requires participants to interact more often and more intensely. More or less interaction is required depending on the experience levels of participants. Novices may need to interact face-to-face and one-on-one when experienced IPT members do not.

Thompson (1967) proposed a contingent relationship between the informational and technological integration requirements of a task and the level of task interdependence created by the organization of work. In Thompson's scheme, successful completion of a task is more likely if the level of task interdependence is appropriate to the integration requirements. Too much or too little interdependence creates a situation where the successful completion of the task is impeded.

Table 1. Types of Interdependence

Type of Interdependence	*Description*
Pooled A B	Pooled interdependence means that participants depend on each other indirectly through the higher aggregate; when designers work on different components, they may not interact except insofar as the whole assembly succeeds or fails. Pooled interdependence does not require more than superficial interaction.
Sequential A → B	Sequential interdependence is asymmetrical, so that A depends on B but not the reverse; an example would be a design waterfall, where the design engineer hands off a part to a manufacturing engineer. Sequential interdependence requires interaction between participants responsible for adjacent functions in the work process.
Reciprocal A ↔ B	Reciprocal interdependence means that the work of each participant interpenetrates the work of the other. It is what we normally mean by collaboration, where the work of each participant directly and immediately influences the work of the other. It requires participants to interact closely.

Source: Adapted from Thompson (1967).

Table 2. Evaluating IPT Capabiltiy Formation Potential

Integration Requirements			
High (Project is Break-through, New Basic Technology or Architecture)	*Type One* Poor primary outcomes in formation, average when mature. Least potential for capability development.	*Type One* Average primary outcomes in formation, good when mature. Moderate potential for capability development.	***Type Two*** **Strong primary outcomes with mix of levels. Most potential for capability development.**
Moderate (Project is New Component or Process)	*Type One* Average primary outcomes in formation, good when mature. Low potential for capability development.	***Type Two*** **Strong primary outcomes with mix of levels. Moderate potential for capability development.**	*Type Three* Good primary outcomes in formation, average when mature. High potential for capability development.
Low (Project is Derivative or Incremental Change)	***Type Two*** **Strong primary outcomes with mix of levels. Moderate potential for capability development.**	*Type Three* Good primary outcomes in formation, average when mature. Moderate potential for capability development.	*Type Three* Average primary outcomes in formation, poor when mature. Moderate potential for capability development.
Interdependence	Pooled	Pooled and Sequential	Pooled, Sequential, and Reciprocal
		Organization of Work	

Using reciprocal interdependence to develop complex new products should be more successful than using sequential interdependence.

If primary outcomes were the only criterion for evaluating a task, we would not need to extend this framework. As our exploratory fieldwork made clear, however, the relationship between the integration requirements of a project and the organization of work is contingent, not causal. In the field we observed that the integration requirements of projects were often disconnected from the levels of task interdependence and types of interaction visible in the organization of the team's work. Team leaders tended to organize the work of the team according to their own preferences and beliefs rather than adhering to the prescriptive theory we brought with us as observers. While we observed only one gross misfit between the integration requirements of the project and the organization of work in the IPT, some of the teams we observed exhibited more task interdependence and interaction than the task appeared to call for. The key observation from our fieldwork was that these tended to be the teams in which participant learning of crossbar skills was the highest.

Our conclusion was that some team leaders organized the work of the IPT to *enhance* interaction between participants and facilitate participant learning of crossbar skills. Other team leaders sought to *reduce* interaction and facilitate primary outcomes, while most team leaders simply *emphasized* the interaction that was appropriate to the project. Once we realized that integration requirements are only loosely coupled to the organization of work, we were able to begin to develop an evaluative framework that let us assess the capability formation potential of IPTs. Because this framework is theory based, it permits us to predict the outcomes of IPTs with characteristics we did not observe in our exploratory fieldwork.

EVALUATING IPT CAPABILITY FORMATION POTENTIAL

The organization of work is the most critical factor for understanding the capability formation potential of IPTs, because it controls the level of participant learning of crossbar skills. For the purpose of qualitative evaluation, we suggest that managers ask themselves whether the organization of work in an IPT reduces, emphasizes, or enhances interdependence and interaction among participants compared to the optimal levels required to achieve successful primary outcomes. For convenience we categorize IPTs as type one, two, or three, in which the organization of work respectively reduces, emphasizes, or enhances interdependence and interaction among participants. Since participant learning is only one element in capability formation, we cannot ignore the basic contingency framework proposed by Thompson because of its connection to primary outcomes. Combining the two creates the contingency table shown in Table 2.

It should be immediately apparent that the off-diagonal normal text in Table 2 represents misfits between project integration requirements and the organization of work. These represent type one IPTs above the diagonal and type three IPTs below. The bold main diagonal represents the range of type two IPTs. IPT type does not relate directly to capability formation. Instead capability formation potential is highest when opportunities for learning all types of T-shaped skills are present and fit between integration requirements and the organization of work leads to desirable primary outcomes.

IPT type is most useful for distinguishing between work organization structures that would typically be used when a strategic capability is mature (types one and two), and structures that would be used when a capability is in the formation stage (types two and three). In organizations like Toyota, where the product development capability is very mature, and projects tend to be driven by existing competencies, projects will tend to be less novel, radical, and complex for the IPTs that work on them, so that level one teams are appropriate for many projects (Ward et al., 1995). Even in an organization like Toyota, however, radically new and complex projects call for type two IPTs designed to create intensive interaction.

This framework can also be applied according to the experience level of IPT participants. IPTs made up of very experienced participants may require less intense interaction to achieve desirable results, even when the project is novel or complex. The reverse is also true: IPTs made up of inexperienced participants may require more interaction to achieve desirable results even when the project is simple and familiar. Therefore Table 2 includes alternative predictions of primary outcomes according to the maturity of the organization or the average experience of the team.

It is difficult for anyone, insider or outsider, to evaluate an ongoing IPT. IPTs do not happen only in public spaces (McDermott, 1995) like meetings—they also exist in corridors and offices, over e-mail, and in participants' imaginations. They are thus hard to observe while they are working. The alternative of waiting until primary outcomes have developed and learning has occurred (or not) is not attractive, because it means giving up an opportunity to intervene. The function of Table 2 is to give managers a framework to use in evaluating IPTs. In order to evaluate the capability formation potential of IPTs, managers need to (1) understand the types of IPTs that are appropriate for the maturity level of the organization and the experience level of the participants, and (2) look beyond nominal adherence to IPT practices, like multidisciplinary membership or concurrent engineering, and see how work is actually organized in practice. These same criteria are also useful in designing and deploying IPTs. In the paragraphs that follow we provide descriptions of IPTs of different types as we encountered them in our fieldwork. It should be noted that all of our examples are drawn from organizations that are still in the formation stage of capability development.

In one IPT we studied, the team leader abdicated responsibility for the project design and the task decomposition. The result was that work was organized by the team into sub-tasks that were linked only by agreement about a common architecture. Participants were able to work more or less alone on subtasks until the various components of the project needed to be assembled. They interacted primarily in weekly (or less frequent) meetings. The project was designed to enact pooled interdependence when in our opinion (and that of many participants) sequential interdependence would have been appropriate. When components did not fit together, participants resorted to accusations and conflict rather than trying to understand the problems and work them out together. These problems were exacerbated by high levels of turnover among participants, including the original team leader. Eventually the new team leader had to resolve problems by fiat. This project came in under budget, on time, and was successful in terms of its products, but participants were quick to point out that it wasted considerably more resources than it should have. It was also clear from participant comments that it created relatively few opportunities for participants to learn T-shaped skills.

This example is characteristic of type one IPTs in an organization in the process of forming capabilities. We would place it in the central cell of the left-hand column in Table 2. When work is organized to reduce interdependence between subtasks, interactions are reduced and learning is impeded. To be successful in primary outcomes, the project should be centrally controlled, with interactions primarily limited to those between the team leader and individual team members. This form of organization requires commitment to a strong plan or architecture, and resistance to deviation from the predetermined project path. When subtasks are linked by a project plan or system architecture, the primary goals of desirable project outcomes can be met, but the IPT provides relatively little opportunity for participant learning other than technical depth. The capability development potential of type one IPTs is low even when primary outcomes are good. They are thus more suitable for organizations with mature capabilities and experienced team participants.

The most common organization of work we observed was moderate reciprocal interdependence within the team and sequential interdependence between the team and the rest of the organization. In some cases the team leader organized work to emphasize the interdependence of team subtasks, while in others an emphasis on interdependence emerged from the team process of decomposing the task. Participants often reported a conscious recognition that their work could be organized to reduce interdependence and interaction, but believed that both they and the project would benefit from recognizing and acting on interdependencies from the outset. This meant actively seeking out other participants to discuss the impact of events in one subtask on other parts of the project, being open and objective about progress and problems, and conforming subtask goals and performance to the project as a whole. It was notable that even when such teams were not colocated, most interaction took place outside of formal spaces like meetings (McDermott, 1995).

Sequential interdependence frequently characterizes the IPT's relationships with parts of the line organization not represented on the team. An example was the relationship between several of the IPTs we studied with the purchasing department, or the production department, or both. Work was often cycled back and forth from team to department, just as in a "siloed" design process, until the department's requirements were met or senior management intervened. In these cases, opportunities for interaction with departments were lost, which reduced the capability formation potential of the IPT.

IPTs with moderate levels of interaction are characteristic of the type two teams we observed in the field. These teams tended to belong to the upper right-hand cell in Table 2, although some (those with more pronounced sequential interdependence) would be positioned farther down and left along the main diagonal. To be successful a type two IPT does not need to be as tightly controlled by the leader. Interactions occur between all participants. When work is organized to emphasize interdependence, participation by outsiders is encouraged. This often manifests itself as procedures to keep outside departments informed and to acquire external expertise. When interdependence between subtasks and/or between the team and its environment is emphasized, the primary goals of desirable project outcomes are often facilitated. Turnover among participants was common, but not wholesale. Participation in type two IPTs provided opportunities for individual development of depth, breadth, integrative, and know-who skills. Type two IPTs sometimes provide opportunities to develop inclusive social skills, as well, but these are likely to be hindered if the team has adversarial relationships with outsiders, since this fosters an "us versus them" mentality. Type two IPTs therefore have considerable capability development potential, and are appropriate to organizations with mature or partially developed capabilities. They are very useful for facilitating learning of T-shaped skills by novices.

We observed several IPTs in which work was organized to enhance interdependence. In these cases the team leader encouraged participants to open subtasks to participation from the whole group. This is sometimes expressed as frequent internal progress and design reviews, but more often seemed to be expressed in the way values and beliefs guided the work. An example was one team's decision to switch from a composite to a metallic material in a key component. The idea came from an external source, and was initially advocated by a team member working on a different component. The whole group engaged in debate about the merits of the alternatives and jointly agreed to the change. The team leader and the component designer both recognized that the decision could have been made without involving other participants, but felt that to do so would not have been legitimate, given the way work was organized. In this and other type three teams we also found that team members were far more likely to work outside of their own areas, and to enjoy doing so. This was often associated with imbalance in the timing of work loads between parts of the project. The flexibility obtained by being able to assign participants outside of their home areas provides an attractive alternative to the more common solution of adding more participants when the project schedule is threatened.

Enhancing interdependence is characteristic of type three IPTs. Like type two, the project does not need to be centrally controlled to be successful, interactions are between all participants, and commitment to particular solutions is delayed as the team generates alternatives. Advanced IPT practices are common and participation by outsiders is encouraged. Turnover appears to be lower than other types of IPT. Type three IPTs work to integrate the team with the functional organization; they are more likely to have all stakeholders represented on the team, and the most likely to make significant efforts to facilitate the work of outside departments.

When interdependence between subtasks is enhanced, some primary project outcomes are liable to suffer. In particular, type three IPTs will be less likely to reduce product development time, and may have higher product development costs. On the other hand, total product and process development cost may be less, as the IPT's efforts to integrate the functional organization may facilitate the transition to production. Participation in a type three IPT provides opportunities for individual development of depth, breadth, integrative skills, and know-who and, unlike other IPTs, significant opportunities to acquire inclusive social skills. They are thus most appropriate for organizations in the early stages of IPT development, and for teams with a high proportion of novices.

CONLUSIONS AND IMPLICATIONS

In this chapter we have argued for the importance of participant learning in integrated project teams to the *formation* of strategic capabilities. We propose that success in the primary outcomes of projects *and* participant learning are both necessary for capability formation through IPTs. The model presented suggests several characteristics that managers can use to qualitatively evaluate the capability formation potential of IPTs. Applying this framework gives managers an opportunity to intervene in IPTs to facilitate capability formation at a level appropriate to the maturity level of the organizational capability.

The best time to intervene in an IPT is before it is formed. The framework we have developed can be used to guide the initial design of IPTs, as well as to evaluate intact teams and intervene in them. It seems obvious to us, as it has to our informants in the field study, that the most effective place to intervene in a team is with the team leader. Team leaders often control the mix of participants and the organization of work, so that choosing team leaders who believe in and will stand up for participant learning is critical for capability formation. We don't propose that team leaders be chosen on this basis alone, since it is equally important that IPTs, particularly new product development IPTs, produce successful primary outcomes. Team leaders need to create the type of team required by three factors:

(1) the maturity of the organizational capability, (2) the experience levels of team participants, and (3) the novelty and complexity of the project.

Our theory has implications for the acquisition of new technology. If it is essential that a radically new technology be acquired for future capability development, then a type three IPT should be designed. The complexity of project objectives should be reduced, so that only moderate levels of integration are required for success. Work should be designed to enhance interaction and place an emphasis on participant learning over primary project outcomes, especially schedule objectives. This will help in overcoming "not invented here" problems and facilitate the integration of the new technology into the existing capabilities of the organization. Follow-on projects should be designed to create type two IPTs that meet both primary and secondary objectives.

Our theory also has implications for the use of virtual teams. While we do not, in principle, believe that virtual teams radically reduce the potential for intense interaction, the use of enabling technology adds an additional layer of mediation between participants. Adding a mediating layer suggests that other sources of complexity should be reduced for the project to be successful. When the enabling technology is difficult to use or unfamiliar, it complicates interaction and participant acquisition of breadth of expertise (Davenport, 1993). Virtual teams also reduce the potential for bodily experience—for the firsthand experience of other people's problems and perspectives (Nonaka, 1994). This suggests that virtual teams may contribute more to capability formation when: (1) the projects are relatively less complex and novel, (2) when the virtual technology is familiar and transparent, and (3) when its use is combined with occasional site visits and face-to-face interactions for novice participants.

This study also has implications for future theory development, as well. The strongest implication is that integrated project teams need to be studied as evolutionary rather than as independent entities. That is, project teams need to be studied as they relate to capability development over the life of several teams. This is different from the argument put forth by Donnellon (1993) and echoed by Dougherty (1996) that teams need to be developed and studied in the context of the organization. What we suggest is that project team researchers follow the path suggested by the strategic project portfolio approach (Wheelwright & Clark, 1992) and examine the relationships between teams past, present, and future. We believe that this means studying the career trajectories of experienced team members through a progression of teams, and linking this to the transmission and diffusion of IPT practices and T-shaped skills.

We have not included motivation in this model, which seems a likely avenue for further developing our theory. In our fieldwork, we found that team members currently participating in projects were usually very highly motivated, even though very few expected to receive any extrinsic reward for participating in a successful project. Most reported that they were highly motivated, because in the IPT environment "we get to do it the way it should be done." Team members who discussed completed projects were often very bitter that they had been put back into the line organization, and that their new skills were not valued or used. One team leader compared his former team members to bacteria which the "white corpuscles of the organization" had eaten up. Team members we studied seldom viewed team participation as "good for your career." The failure to reassign team members is a distinct problem from the one we address here; it has many possible causes which are beyond the scope of this chapter, but clearly belong to the question of capability formation.

A final direction for future research is to more closely examine the relationship between various practices employed in IPTs and diversity of membership. A variety of authors (Hitt et al., 1993; Nonaka, 1994; Purser, Pasmore, & Tenkasi, 1992) have discussed barriers to and facilitators of participant learning in multidisciplinary teams, but the relationship has not been tested. The question of whether participant learning is the result of practice and diversity jointly or independently is an empirical one that bears directly on the design of IPTs and the formation of strategic capabilities.

REFERENCES

Barley, S.R., & Orr, J.E. (1997). Introduction: The neglected workforce. In S.R. Barley and J.E. Orr (Eds.), *Between craft and science: Technical work in U.S. settings*. Ithaca, NY: Cornell University Press.

Barney, J.B. (1991). Firm resources and sustained competitive advantage. *Journal of Management, 17*(1), 99-120.

Blackler, F. (1995). Knowledge, knowledge work and organizations: An overview and interpretation. *Organization Studies, 16*(6), 1021-1046.

Boland, R.J., Jr., & Tenkasi, R.V. (1995). Perspective making and perspective taking in communities of knowing. *Organization Science, 6*(4), 350-372.

Bowen, H.K., Clark, K.B., Holloway, C.A., & Wheelwright, S.C. (1994a, September-October). Make projects the school for leaders. *Harvard Business Review, 71*, 131- 140.

Bowen, H.K., Clark, K.B., Holloway, C.A., & Wheelwright, S.C. (1994b, September-October). Development projects: The engine of renewal. *Harvard Business Review, 71*, 110-120.

Brown, J., & Duguid, P. (1991). Organizational learning and communities-of-practice: Toward a unified view of working, learning and innovation. *Organization Science, 2*(1), 40-57.

Cohen, W.M., & Levinthal, D.A. (1990) . Absorptive capacity: A new perspective on learning and innovation. *Administrative Science Quarterly, 35*, 128-152.

Collis, D.J., & Montgomery, C.A. (1995, July-August). Competing on resources: Strategy in the 1990s. *Harvard Business Review, 72*, 118-129.

Conner, K.R. (1991). A historical comparison of resource-based theory and five schools of thought within industrial organization economics: Do we have a new theory of the firm? *Journal of Management, 17*(1), 121-154.

Cummings, L.L. (1977). Emergence of the instrumental organization. In P.S. Goodman, J.M. Pennings, and Associates (Eds.), *New perspectives on organizational effectiveness* (pp. 56-62). San Francisco: Jossey-Bass.

Davenport, T.H. (1993). *Process innovation: Reengineering work through technology*. Boston: Harvard Business School Press.

Department of Defense. (1996). *DoD guide to integrated product and process development, Version 1.0.* Washington, DC: Acquisition and Technology, Office of the Under Secretary of Defense.

Donnellon, A. (1993). Cross-functional teams in product development: Accommodating the structure to process. *Journal of Product Innovation Management, 10,* 377-392.

Dougherty, D. (1992). Interpretive barriers to successful product innovation in large firms. *Organization Science, 3*(2), 179-202.

Dougherty, D. (1996). Organizing for innovation. In S.R. Clegg, C. Hardy, & W.R. Nord (Eds.), *Handbook of organization studies* (pp. 424-439). Newbury Park, CA: Sage.

Dougherty, D., & Hardy, C. (1996). Sustained product innovation in large, mature organizations: Overcoming innovation-to-organization problems. *Academy of Management Journal, 39*(5), 1120-1153.

Dougherty, D., & Heller, T. (1994). The illegitimacy of successful product innovation in established firms. *Organization Science, 5*(2), 200-218.

Glynn, M.A., Lant, T.K., & Miliken, F.J. (1994). Mapping learning processes in organizations: A multi-level framework linking learning and organizing. In C. Stubbart, J. Meinkl, & J. Porac (Eds.), *Advances in Managerial Cognition and Organizational Information Processing, 5,* 43-83.

Goodman, P.S., & Leyden, D.P. (1991). Familiarity and group productivity. *Journal of Applied Psychology, 76*(4), 578-586.

Harris, D.M., & DeSimone, R.L. (1994). *Human Resource Development.* Fort Worth: The Dryden Press.

Henderson, R.M. (1996). Technological change and the management of architectural knowledge. In M.D. Cohen & L.S. Sproull (Eds.), *Organizational Learning.* Newbury Park, CA: Sage.

Hitt, M.A., Hoskisson, R.E., & Nixon, R.D. (1993). A mid-range theory of interfunctional integration, it antecedents and outcomes. *Journal of Engineering and Technology Management, 10,* 161-185.

Hitt, M.A., Ireland, R.D., & Hoskisson, R.E. (1997). *Strategic management: Competitiveness and globalization.* Minneapolis, MN: West Publishing Company.

Iansiti, M. (1993, May-June). Real world R&D: Jumping the product generation gap. *Harvard Business Review, 74,* 138-147.

Jackson, S.E., May, K.E., & Whitney, K. (1995). Understanding the dynamics of diversity in decision making teams. In R.A. Guzzo & E. Salas (Eds.), *Team effectiveness and decision making in organizations* (pp. 204-261). San Francisco: Jossey-Bass.

Katzenbach, J.R., & Smith, D.K. (1994) . *The wisdom of teams: Creating the high performance organization.* New York: HarperBusiness.

Klein, J.A. (1994). Maintaining expertise in multi-skilled teams. In M.M. Beyerlein, D.A. Johnson (Eds.), *Advances in interdisciplinary studies of work teams: Vol. 1. Theories of self-managing work teams* (pp. 145-165).

Klein, J.A. & Maurer, P.M. (1995). Integrators not generalists needed: A case study of integrated product development teams. In M.M. Beyerlein, D.A. Johnson, & S.T. Beyerlein (Eds.), *Advances in interdisciplinary studies of work teams: Vol. 2. Knowledge work in teams* (pp. 93-116).

Lave, J., & Wenger, E. (1991). *Situated learning: Legitimate peripheral participation.* New York: Cambridge University Press.

Lawler, E.E., III, (1992). *The ultimate advantage: Creating the high involvement organization.* San Francisco: Jossey-Bass.

Leonard-Barton, D. (1992). Core capabilities and core rigidities: A paradox in managing new product development. *Strategic Management Journal, 13,* 115-125.

Leonard-Barton, D. (1995). *Wellsprings of knowledge: Building and sustaining the sources of innovation.* Boston: Harvard Business School Press.

Manz, C.C., & Sims, H.P. (1993). *Business without bosses: How self managing teams are building high performing companies.* New York: Wiley.

McDermott, R. (1995). Working in public—learning in action: Designing collaborative knowledge work teams. In M.M. Beyerlein, D.A. Johnson, & S.T. Beyerlein (Eds.), *Advances in interdisciplinary studies of work teams: Vol. 2. Knowledge work in teams* (pp. 35-60).

Mohrman, S.A., Mohrman, A.M., Jr. & Cohen, S.G. (1995). Organizing knowledge work systems. In M.M. Beyerlein, D.A. Johnson, & S.T. Beyerlein (Eds.), *Advances in interdisciplinary studies of work teams: Vol. 2. Knowledge work in teams* (pp. 61-92).

Moreland, R.L., Argote, L., & Krishnan, R. (1996). Socially shared cognition at work: Transactive memory and group performance. In J.L. Nye & A.M. Brower (Eds.), *What's social about social cognition? Research on socially shared cognition in small groups* (pp. 57-84). Newbury Park, CA: Sage.

Nonaka, I. (1994). A dynamic theory of organizational knowledge creation. *Organization Science, 5*(1), 14-37.

Northcraft, G.B., Polzer, J.T., Neale, M.A., & Kramer, R.M. (1996). Diversity, social identity and emergent social dynamics in cross-functional teams. In S.E. Jackson & M.N. Ruderman (Eds.), *Diversity in work teams: Research paradigms for a changing workplace* (pp. 85-97). Washington, DC: American Psychological Association.

Orr, J. (1990). Sharing knowledge, celebrating identity: Community memory in a service culture. In D. Middleton & D. Edwards (Eds.), *Collective remembering* (pp. 61-78). London: Sage.

Orsburn, J., Moran, L., Musselwhite, E., Zenger, J. & Perrin, C. (1990). *Self-directed work teams: The new American challenge*. Homewood, IL: Business One Irwin.

Peteraf, M.A. (1993). The cornerstones of competitive advantage: A resource based view. *Strategic Management Journal, 14,* 179-191.

Prahalad, C.K., & Hamel, G. (1990, May-June). The core competence of the corporation. *Harvard Business Review, 69*, 79-91.

Purser, R.E., & Montuori, A. (1995). Varieties of knowledge work experience: A critical systems inquiry into the epistemologies and mindscapes of knowledge production. In M.M. Beyerlein, D.A. Johnson, & S.T. Beyerlein (Eds.), *Advances in interdisciplinary studies of work teams: Vol. 2. Knowledge work in teams* (pp. 117-162).

Purser, R.E., Pasmore, W.A., & Tenkasi, R.V. (1992). The influence of deliberations on learning in new product development teams. *Journal of Engineering and Technology Management, 9,* 1-28.

Rulke, D.L., & Rau, D. (1997). Examining the encoding process of transactive memory in group training. In L.N. Dosier & J.B. Keys (Eds.), *Academy of management best paper proceedings* (pp. 349-353).

Shapiro, D.L. (1993). Reconciling theoretical differences among procedural justice researchers by re-evaluating what it means to have one's views 'considered': Implications for third party managers. In R. Cropanzano (Ed.), *Justice in the workplace: Approaching fairness in human resource management* (pp. 51-78). Hillsdale, NJ: L. Erlbaum Associates.

Sims, H.P. & Manz, C.C. (1994). The leadership of self-managing work teams. In M.M. Beyerlein, D.A. Johnson (Eds.), *Advances in interdisciplinary studies of work teams: Vol. 1. Theories of self-managing work teams* (pp. 187-222).

Stasser, G., Stewart, D.D., & Wittenbaum, G.M. (1995). Expert roles and information exchange during discussion: The important of knowing who knows what. *Journal of Experimental Social Psychology, 31,* 244-265.

Tatikonda, M.V. (1995). *Technology planning and implementation in product development projects: An empirical study of innovation type, organization, and performance*. Unpublished doctoral dissertation, Graduate School of Management, Boston University.

Thompson, J.D. (1967). *Organizations in action*. New York: McGraw-Hill.

Tjosvold, D., & Tjosvold, M.M. (1995). Cross-functional teamwork: The challenge of involving professionals. In M.M. Beyerlein, D.A. Johnson, & S.T. Beyerlein (Eds.), *Advances in interdisciplinary studies of work teams: Vol. 2. Knowledge work in teams* (pp. 1-34).

Tushman, M.L., & O'Reilly, C.A., III. (1997). *Winning through innovation : A practical guide to leading organizational change and renewal.* Boston, MA: Harvard Business School Press.

Wageman, R. (1995). Interdependence and group effectiveness. *Administrative Science Quarterly, 40,* 145-180.

Ward, A., Liker, J.K., Cristiano, J.J., & Sobek, D.K., II. (1995, Spring). The second Toyota paradox: How delaying decisions can make better cars faster. *Sloan Management Review, 36*, 43-61.

Wegner, D.M. (1987). Transactive memory: A contemporary analysis of the group mind. In B. Mullen & G.R. Goethals (Eds.), *Theories of group behavior* (pp. 185-208). New York: Springer-Verlag.

Wheelwright, S.C. & Clark, K.B. (1992, March-April). Creating project plans to focus product development. *Harvard Business Review, 72*, 70-82.

APPLYING BEHAVIOR MODIFICATION TO THE MANAGEMENT OF TEAM PERFORMANCE

A SYSTEMS PERSPECTIVE

Michael T. Barriere, Ira T. Kaplan, and William Metlay

ABSTRACT

Given the importance of teams in today's business environment, methods need to be developed for maximizing team performance. Traditional management practices are often based on techniques that encourage individual productivity at the expense of team performance. Extending performance management to teams requires the development of a comprehensive framework to address the full range of variables associated with group behavior. To develop such a framework, this chapter analyzes the variables investigated in empirical studies of team performance and classifies them within a systems model of work group behavior. This analysis demonstrates the range of variables that need to be considered when the techniques of performance management are extrapolated from individuals to groups. Specifically, many variables that are not relevant to individual performance are shown to be essential to the design and evaluation of team-based interventions.

Advances in Interdisciplinary Studies of Work Teams, Volume 6, pages 117-137.

ISBN: 0-7623-0655-6

INTRODUCTION

The foundation for performance management is Skinner's (1938, 1953) research on operant conditioning. Skinner's operant model holds that behavior is a function of its antecedents and its consequences. Antecedents are stimuli that set the stage for a response to occur. Consequences are stimuli that follow the response and influence the probability of its subsequent occurrence. In the operant model, consequences are classified according to their effects. Three types of consequences are positive reinforcement, negative reinforcement, and punishment. Both positive and negative reinforcement increase the likelihood of a response. Positive reinforcement is the presentation of a stimulus which increases the likelihood of the behavior that it follows. Negative reinforcement is the removal of a stimulus which increases the likelihood of the behavior that precedes its removal. In contrast, punishment decreases the likelihood of behavior that it follows. According to the operant model, antecedents and consequences are the principal determinants of behavior.

The application of the operant model to behavior analysis in the work place is referred to as organizational behavior modification (OB Mod), organizational behavior management (OBM), and performance management (PM). All three terms are used more or less interchangeably in the literature. Daniels (1989) defined performance management as a systematic, data-oriented approach to managing people at work that relies on positive reinforcement as the major way to maximize performance. This approach is implemented through an ABC model of antecedents (A), behaviors (B), and consequences (C), which identifies task-related target behaviors and relates them to specific antecedents and consequences, using frequent, objective, and quantitative measurement. In the workplace, antecedents include performance standards, verbal instructions, goals, and procedures. Behavior is what the employee does as part of the task. Consequences are events contingent upon targeted task behaviors, including recognition, cash bonuses, plaques, and time off with pay.

MAXIMIZING TEAM PERFORMANCE

Until recently, performance management has focused on maximizing individual performance, rather than the performance of teams. Because of its origin in the analysis of individual behavior, performance management has traditionally emphasized individual antecedents, behaviors, and consequences. Justifying this emphasis, Daniels (1994) points out that reinforcement is highly individualistic: A consequence that is a positive reinforcer for one person might not be for another. A similar argument can be made for antecedents. Therefore, since it is not practical to optimize antecedents and consequences for an entire team, Daniels advocates focusing on the behaviors of individual team members. There is evidence, however, that maximizing individual performance may not maximize the performance of teams. When work is highly interdependent, striving for individual goals or rewards can undermine team performance by encouraging competition at the expense of cooperation (Hitchcock, 1990; Johnson & Johnson, 1989; Mitchell & Silver, 1990; Mohrman, Mohrman, & Lawler, 1992).

Competition Within Teams Can Be Counterproductive

Research shows that competition is counterproductive when team members must cooperate to accomplish their work. Both competition and cooperation can be encouraged by the manipulation of either goals (antecedents) or rewards (consequences). In a laboratory experiment related to goals, Mitchell and Silver (1990) gave three-person groups the task of building a tower with wooden blocks under different goal conditions. In the individual goal condition, each subject's goal was to get at least 7 blocks on the tower in 15 seconds. In the group goal condition, the subjects were told to get at least 21 blocks on the tower in 15 seconds. The groups that were given a group goal collaborated with one another and consequently performed better than the groups whose members were given competitive goals.

Research involving the manipulation of rewards produced similar results. Deutsch (1949) had students work on group projects under reward conditions designed to stimulate either competition or cooperation within the group. In the competitive condition, the students were graded on their individual contribution to their group project. In the cooperative condition, students received a grade based on their group's overall performance. He found that the cooperative groups produced projects of higher quantity and quality than the students who were rewarded individually. Other studies have also shown that when a task requires cooperation, group rewards result in higher performance than individual rewards (Johnson & Johnson, 1989; Johnson, Maruyama, Johnson, Nelson, & Skon, 1981; Miller & Hamblin, 1963).

Extending Performance Management from Individuals to Teams

Traditionally, organizations have reinforced individual rather than team performance, through individual rewards (e.g., salary raises and bonuses), through selection decisions based on individual skills rather than work team composition, and through the emphasis on individual goals as part of the yearly performance appraisal process. Given the importance of teams in today's business environment (Druckman, Singer, & Van Cott, 1997; Mintzberg, 1979), methods need to be developed for maximizing team performance. Unfortunately, maximizing the performance of teams requires more than the predominant, individually oriented performance management practices of most organizations (Mohrman et al., 1992). Extending performance management to teams requires the development of a comprehensive framework that addresses the full range of variables associated with

Table 1. Illustrative Variables for the Systems Model of Work Group Behavior

	Stage			
Variable Type	*Input*	*Process*	*Output*	*Feedback*
Task	Complexity Variety Goal clarity	Problem solving	Quality of product Quantity of product	Knowledge of job results
Individual	Abilities Knowledge	Participation	Satisfaction	Superior-subordinate annual review
Group	Group size Group composition	Intragroup interaction	Changes in member relations	Teamwork discussions
Environment	Organization policies Reward systems	Intergroup cooperation	Effects on other groups	Management evaluation of the group's contribution to the organization

Source: Adapted from Metlay, Kaplan, & Rogers (1994).

group performance. To help develop such a framework, the present chapter analyzes the variables investigated in research on team performance management and classifies them within a systems model of work group behavior.

Systems Model of Work Group Behavior

The model chosen for extending performance management from individuals to teams, is one that has previously been applied to the analysis of quality circles (Greenbaum, Kaplan, & Metlay, 1988), team-based organizational development interventions (Kaplan & Greenbaum, 1991), and self-managed work teams (Metlay, Kaplan, & Rogers, 1994). This model classifies the variables manipulated and measured in research on work group behavior into four stages: input, process, output, and feedback. Input variables are antecedent conditions that exist before any group behavior is observed. They include such variables as specification of the group's task, individual member characteristics, group size and composition, and organizational policies. Process variables are events, activities, and individual and group behaviors such as problem solving, communication, and cooperation, which occur over time and are influenced by the input variables. Output variables are the results of processes. They include measures of task outcomes such as quality and quantity, individual outcomes such as job satisfaction, group outcomes such as cohesiveness, and environmental outcomes such as customer satisfaction and the group's reputation. Finally, feedback variables comprise information about processes or outputs that are provided to members of the work group. In experimental research and in organizational interventions, input and feedback variables are often the independent variables manipulated by researchers, consultants, or managers, whereas processes and outputs are the dependent measures. Table 1 provides illustrative examples for the variables that comprise each cell of the model.

The purpose of this chapter is to analyze the antecedents, behaviors, and consequences investigated in empirical studies of team performance and classify them within the more refined categories of the systems model of work group behavior. Specifically, our purpose is to determine the range of variables that need to be considered when the techniques of performance management are extrapolated from individuals to groups and to discover the extent to which variables not relevant to individual performance are essential to the design and evaluation of team-based interventions.

METHOD

Selection of Studies

A literature search was conducted to identify empirical research studies that investigated the effects of antecedents and consequences on work group behavior.

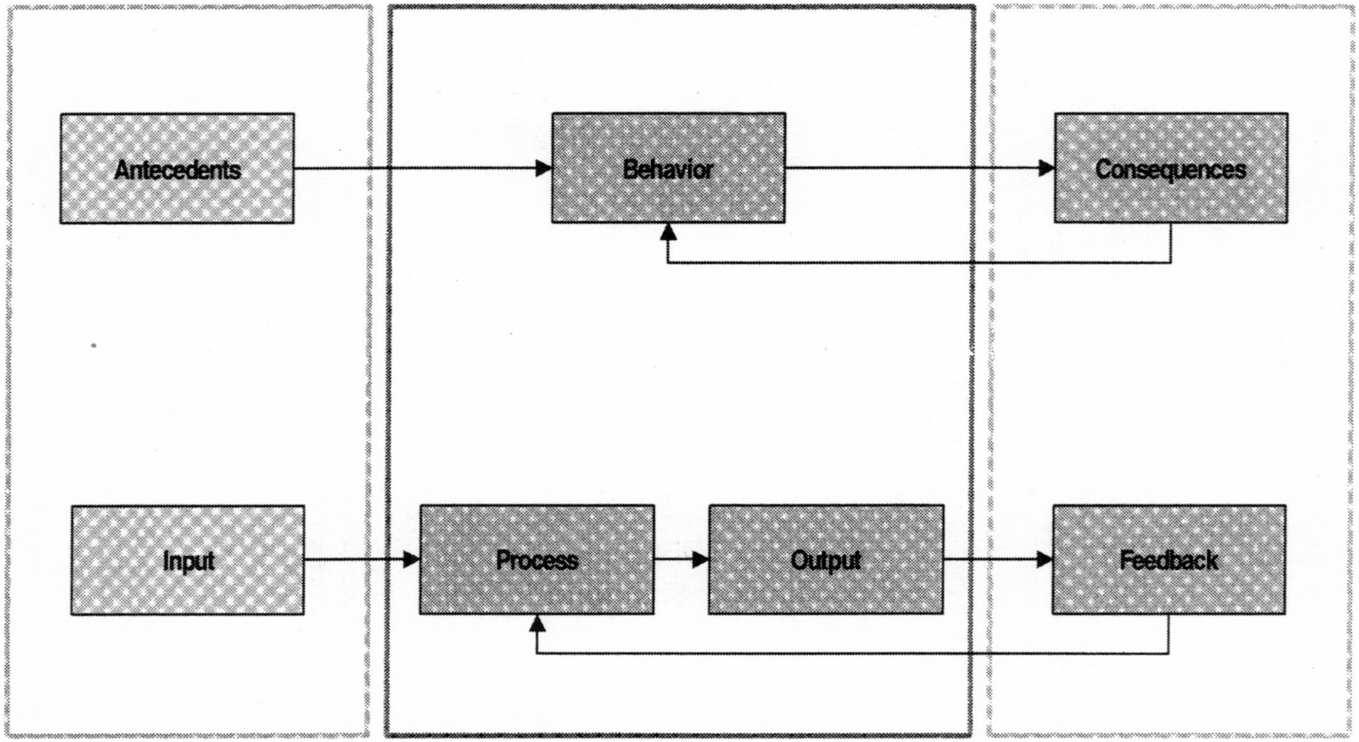

Antecedents

Behavior

Consequences

Input

Process

Output

Feedback

Figure 1. Relationship Between the ABC Model and the Systems Model

To find such studies, a computer-based search of PsycLit for the period 1980-1997 was conducted using the key words *performance management*, *behavior management*, *behavior modification*, *groups*, *teams*, *feedback*, and *reinforcement.* The studies were selected to satisfy two criteria:

1. They contained variables central to performance management.
2. Together as a set they represented all the categories of variables identified by the systems model of work group behavior.

Relationship Between the ABC Model and the Systems Model

The four stages of the systems model are related to the antecedents, behaviors, and consequences of the ABC model of performance management. As shown in Figure 1, antecedents and consequences correspond to inputs and feedback, respectively. On the other hand, behaviors in the ABC model do not distinguish two types of variables which are classified into process and output categories by the systems model. When behavior is an ongoing activity, such as putting merchandise on shelves in a store or building a tower of blocks, it is a process. When behavior is described as a result, such as the percent of shelves that are stocked or the height of the completed tower, it is an output.

When the ABC model of performance management is extended to groups, it is particularly deficient in its ability to distinguish the variety of inputs, processes, outputs, and feedback which are significant in group as opposed to individual behaviors. In contrast, the systems model of work group behavior distinguishes four types of variables in each of its four stages. In each stage there are variables that describe (1) the group's task, (2) each individual group member, (3) group members considered together, and (4) the group's environment. The following paragraphs define the four types of variables in each stage (input, process, output, and feedback).

Task input variables are antecedent conditions that specify the group's task. They include group goals and performance criteria that are presented to the group members before they begin the task or at the beginning of a performance management intervention. *Individual inputs* are characteristics of individual group members, such as knowledge and ability. In performance management interventions, individual inputs include instructions specifically given to an individual. *Group inputs* specify group structure, composition, or size. They include patterns or combinations of individual characteristics and instructions given to more than one individual. *Environmental inputs* are variables outside the group that may affect it, such as an organization's reward policy.

Task process variables describe performance of the group's task without reference to the behavior of any person or persons. They include variables that describe the procedure being implemented or the rate at which the task is being performed. *Individual process* variables describe the behavior of a single individual as it occurs over time, such as time spent on one activity or another. *Group process* variables describe interactions between members of the group, whereas *environmental process* variables describe interactions with the environment, such as competition between groups.

Task output variables describe the completed task, such as quality and quantity of a finished product. *Individual outputs* are characteristics of an individual or results attributable to an individual after the task is completed, such as satisfaction with the work or an intention to continue working on the task. *Group Outputs* are characteristics or results for several individuals or the entire group, such as group morale or a shared intention to continue working as a group. *Environmental Output* variables include measures of the group's effects on its environment, such as customer satisfaction, and relationships between the group's performance and that of other groups.

Feedback variables provide information about processes and outputs. Consistent with this definition, reinforcers are classified as feedback variables. Different types of feedback are classified on the basis of their content. Thus, *task feedback* provides information about, or is contingent on, the task. *Individual feedback* provides information about, or is contingent on, an individual's behavior or accomplishment. Similarly, *group feedback* provides information about, or is contingent on, the group process or output; and *environmental feedback* refers to environmental processes or output.

PROCEDURE

The independent and dependent variables described in each of the selected performance management studies were classified into the cells of the group behavior model, as defined above. Each study was thus summarized in terms of the cells that were investigated and the relationships that were found between those cells. The number and type of cells investigated in each study were then compared to the full 16-cell model, to obtain a measure of that study's comprehensiveness or completeness. Finally, the results were aggregated to provide an overview of how often the various cells of the systems model were investigated in this particular set of studies.

RESULTS

The literature search identified five studies which extended the general performance management paradigm to groups, and which together contain variables that represented all 16 cells of the group behavior model. Three of these studies were conducted in field (real-world) settings and had a behavioral orientation. They were described by their authors as practical applications of "behavior management" (Luthans, Paul, & Baker, 1981), "an operant model" (Komaki, Desselles, & Bowman, 1989), and "organizational behavior management" (Kortick & O'Brien, 1996). The other two studies used student participants and were based on cognitive models, goal setting (Mitchell & Silver, 1990), and equity theory (Barr & Conlon, 1994). However, they manipulated variables, that is, group goals and feedback, which are central to the performance management of teams.

The five studies are presented below in chronological order. The method and results of each study are summarized, and the variables manipulated and measured are classified into the corresponding categories of the 16-cell model. Within each summary, the formal terminology of the model is placed in brackets to differentiate it from the language used in the original paper.

Improving Salespersons' Performance by Contingent Reinforcement

Luthans, Paul, and Baker (1981) studied the impact of contingent reinforcement on salespersons' performance. They called their approach behavior management, and said that it was based on operant learning theory, which assumed that employee behavior was a function of environmental contingencies. The salespersons worked in a department store. Eight departments were randomly assigned to an experimental group and eight to a control group. Each department had a manager and five or six salespersons.

Salespersons' performance behaviors were observed twice an hour during each workday. The observed behaviors were classified into five categories: (1) selling, (2) stock work, (3) miscellaneous work-related behavior other than selling or stock work, (4) idle time, such as socializing with co-workers or standing around, and (5) absence from the work station. (In the 16-cell model, these five behaviors are classified in the *individual process* cell). These observations were collected during a four-week baseline period. They continued to be collected during a four-week intervention period, in which the experimental group was reinforced at the end of each week. Finally, they were collected during a four-week post-intervention period, in which the reinforcement was withdrawn from the experimental group. The same five behaviors were observed for the control group, which did not receive reinforcement during the entire 12-week period.

On the basis of the above five categories of behavior, the following performance standards were developed: (1) The salespersons should be present in the department during assigned working hours. (2) Customers should be assisted promptly. (3) The display shelf should be filled to at least 70 percent of its capacity. At the beginning of the intervention, the three performance standards were carefully explained to salespersons in both the experimental and the control groups. (The introduction of these standards is classified in the input stage of the model. Because the first two standards apply to each individual salesperson, they are classified under *individual input*. The third standard however, defines a task goal for the entire group, therefore it is a *task input*). In addition, salespersons in the experimental group were told that they would receive rewards based on how well they met the standards. The weekly reward consisted of time off with pay or equivalent cash, plus an opportunity to compete in a drawing for a company-paid one-week vacation for two. The number of hours off with pay, or equivalent cash, was on a graduated scale, so that, as more of the desired behavior was exhibited the amount of the reward was increased. The drawing was held at the end of the four-week intervention. (At the beginning of the intervention, telling each salesperson what the individual rewards would be is an *individual input*. The actual presentation of these rewards to each salesperson at the end of each week is *individual feedback*. Management's establishment of the reward policy is an *environmental input)*.

There was no difference in the baseline behaviors of the experimental and control groups. On the first day of the intervention, after the experimental participants had been informed of the contingent reinforcement schedule, there was a marked change in their behavior. This change persisted over the four weeks of the intervention and the four-week post-intervention period. The experimental group increased its retailing behavior (selling, stockwork, miscellaneous) and decreased idle time and absence from the workstation, but the control group did not. (Restating these results in terms of the model, the change in the behavior of the experimental participants on the first day of the intervention is an effect of *task input* and *individual input* on *individual process*. The maintenance of this effect over the four-week intervention period is a consequence of *environmental input* and *individual feedback*). During the post-intervention phase, the reward policy (*environmental input*) was discontinued. However, the differential behavior of the two groups persisted, presumably because the more effective behavior of the experimental group was maintained by such natural reinforcers as supervisor's praise and more positive customer reactions (a continued effect of *individual feedback* on *individual process)*.

Extending an Operant Model of Effective Supervision to Teams

Komaki, Desselles, and Bowman (1989) took an operant learning approach to the analysis of group behavior. Their approach differed from Luthans and colleagues'(1981) operant learning approach, in that it was observational rather than experimental, and their group's task was more interdependent. Each group or team consisted of a three-person crew and a skipper whose task was to participate in a sailboat regatta, or series of races. Nineteen skippers participated in a total of six ten-boat races, for an average of 3.2 races per skipper. An observer on each boat recorded the behavior of the skipper during preparation for the race and during the race itself. Preparation began when the skipper set foot on the boat and continued until the three-minute warning whistle before the race began. (In terms of the 16-cell model, specification of the task, i.e., preparation for racing, is a *task input* variable).

Six performance-related, verbal behaviors of the skipper were recorded: antecedents, monitors, and consequences, with and without coordination. These behaviors were defined as follows: (1) individual antecedents, that is, giving instructions to an individual crew member; (2) coordination antecedents, that is, giving instructions that involve interaction; (3) monitoring individual behavior, that is, asking a question of an individual that does not involve coordination with another person; (4) monitoring team coordination, that is, asking for information about an interaction; (5) individual consequences, that is, giving feedback to a crew member about that individual's performance; and (6) coordination consequences, that is, giving feedback about an interaction. (Classifying these variables in cells of the model reveals their complexity. In a general sense, all these behaviors are *individual process*, i.e., verbal behavior of the leader, but they are also inputs and feedback to one or more crew members. Giving instructions to and asking questions of an individual are *individual inputs* to that person. Instructions and questions involving coordination are classified as *group inputs*. Giving feedback about individual performance is *individual feedback*, and feedback about coordination is *group feedback*).

Two measures of skipper effectiveness were used in the research: (1) series standing, calculated from the number of points awarded for finishing at various positions in a race, and (2) coaches' ratings and rankings of each skipper's crew handling performance. (In terms of the model, these measures are both *environmental outputs*, because they express the skipper's performance in relation to the performance of other ships or to the impressions made on other persons [coaches] in the environment. If the time for a skipper to complete a race had been used as a measure, that would have been an *individual output*).

Komaki and colleagues (1989) reported that during the race, skippers spent much more time providing feedback and giving instructions than they did during the preparation phase. (Since preparation vs. racing is a *task input* variable, and providing feedback and giving instructions are *individual process* variables, this result can be described as an effect of *task input* on *individual process*). They also found that racing success was significantly correlated with the frequency of individual monitors and consequences during the race, but not with their frequency during preparation. (This result suggests that *individual input* and *individual feedback* during the race influenced *environmental outputs*). There was no relationship between the frequency of coordination behaviors and outcomes.

Effects of Individual and Group Goals on Interdependent Tasks

Mitchell and Silver (1990) employed a goal-setting model, which reflected a cognitive approach, as opposed to the behavioral orientation of the preceding studies. Nevertheless, their variables can be described in objective, behavioral terms and classified into the cells of our model. Mitchell and Silver (1990) examined the effects of four different goal-setting conditions on the performance of groups working on an interdependent task. The group's task was to build a single tower of blocks. The participants were 96 female psychology students, who made up 32 three-person groups. Eight groups were assigned to each of the following conditions:

1. Individual goal, where participants were told they would receive one point for each block of their color in the tower and that each individual's goal was to get at least 7 blocks of her own color on the tower.
2. Group goal, where the team received a point for each block on the tower, and the goal was to get at least 21 blocks on the tower.
3. Individual goal plus group goal, where both individual and team points were awarded, and each individual's goal was to get at least 7 blocks on the tower, while the group's goal was to get at least 21 blocks on the tower.
4. No specific goal, where the goal was to get as many blocks on the tower as possible.

Each group performed ten 15-second trials. The groups with goals received feedback after each trial about their individual or group performance, as appropriate to their goals. (The goal variable manipulated in this experiment extends over three levels of input: Each individual's goal considered separately is an *individual input*. All three individual goals together comprise a *group input*. The single group goal assigned to the entire team is a *task input*, since the task level of the model refers to the task of the entire group. By the same reasoning, the feedback that is part of each goal condition is classified as *individual feedback*, *group feedback*, and *task feedback*).

After the last trial, the participants were asked to write a description of the strategy they developed for putting blocks on the tower. Two raters reviewed all the strategies and agreed on their classification as cooperative or competitive. Cooperative strategies included such behaviors as going slowly, taking turns, and balancing the blocks to avoid knocking them over. In contrast, competitive strategies included trying to go first and putting blocks on quickly without regard for others. (In terms of the model, these strategies are *group process*). The researchers also recorded a measure of ongoing activity that referred to the construction of the tower as opposed to the behavior of the participants, that is, the number of times that three or more blocks fell off the tower while it was being built. (Because this measure describes the building of the tower without reference to any human behavior, it is classified as a *task process*). Two other dependent measures were the number of blocks on the tower at the end of the trial (a *task output*) and the percentage of occasions in which each of the three block colors appeared in successive three-block sequences in the completed tower. (Because this alternation of colors was the result of a cooperative strategy, it is classified as a *group output*).

The individual goal condition produced results that were different from the results of the other three conditions. There were no significant differences between the group-goal condition, the group-goal-plus-individual-goal condition, and the no-specific-goal condition. Compared to the other three conditions, the individual-goal condition resulted in less cooperative behavior and more competition, more falls and less turn-taking, and fewer blocks on the tower at the end of the trial. (Thus we have effects of *task, individual, and group inputs* on *task and group processes* and *task and group outputs*.)

Effects of Distribution of Feedback in Work Groups

Barr and Conlon (1994) investigated the effects of several types of feedback on group members' intentions to continue working on a decision-making task. Although their interest in these types of feedback was motivated by considerations of equity theory, their experiment can be described in terms of performance management. Participants were 180 business students, who were randomly assigned to 60 three-person groups. Each group participated in a decision-making simulation at a university. Their task was to allocate a budget and set prices for three markets: mainframe computers, minicomputers, and microcomputers. They were told that their individual and group decisions would be evaluated by a computer model, and that they would receive group and individual feedback after each session. They were also told that they would receive course credit based on their group's performance. Every group received the same feedback for five baseline sessions, after which the feedback was manipulated for four experimental sessions. After the last experimental session, each participant was asked to indicate, on an 11-point scale, his or her willingness to continue participating in the simulation. This response was the dependent variable in the experiment (an *individual output*).

The independent variables were the sign of the feedback to the individual (positive or negative), the distribution of feedback received by members of the group (majority positive or majority negative), and the sign of the group feedback (positive or negative). (In terms of the 16-cell model, feedback to the individual is *individual feedback*, the distribution of feedback received by members of the group is *group feedback*, and feedback about the group's task performance is *task feedback*).

The results showed that all three feedback variables influenced intention to continue working at the task. With respect to feedback about the group's task performance, positive feedback resulted in greater intention to persist than negative feedback (an effect of *task feedback* on *individual output)*. Positive individual feedback produced a greater intention to persist than negative individual feedback (an effect of *individual feedback* on *individual output*). Majority positive feedback produced a greater intention to persist than majority negative feedback (an effect of *group feedback* on *individual output*). In addition to these three main effects, there were also significant interactions. There was a significant two-way interaction between *individual feedback* and *group feedback*. The effect of individual feedback was greater when the majority of the group received positive, rather than negative, feedback. In fact, individual feedback had little effect when the majority of the group received negative feedback. There was also a significant three-way interaction involving all three types of feedback: task, individual, and group.

Effects of Team-Based Feedback and Reinforcement on Team Performance

Kortick and O'Brien (1996), using an organizational behavior management approach, introduced a program of feedback and positive reinforcement that encouraged competition between work groups. The objective of the program was to increase service quality at a shipping facility of a package delivery company. This facility was responsible for distributing packages throughout the country. Packages were sorted according to destination and then loaded on trucks and trailers by 13 work teams of eight employees each.

The program was modeled on baseball, with the teams divided into a National League and an American League. Within each league, pairs of opposing teams played five games a week. (Competition between teams is an *environmental process*. If there were competition within teams, that would have been a *group process*). A game was a daily comparison of performance in loading the trucks and trailers. The score of the game was calculated on the basis of each team's performance on three quality control measures: load quality (how well the packages had been stacked into trucks and trailers on a 1 to 10 scale, where 10 represented a solid wall with no overhanging boxes), load retention (percentage of the load that was strapped down correctly), and missort rate (ratio of the number of outgoing packages loaded on the correct truck or trailer for every package that was loaded incorrectly). Data were collected on these three measures over a 12-week baseline before the program was presented to the employees. (The three quality control measures were *task outputs*).

Following baseline, the program was explained to the employees. They were told that standings would be posted, which would include each team's scores, position in the league, and next opponent. In addition, each team's weekly and monthly average ratings on the three quality-control measures would also be posted on the standings board. Weekly rewards (pizza and beer) were provided for the team with the best overall performance for that week. At the end of the month, the team with the most points was rewarded with plaques and dinner at a nearby restaurant. A play-off and world series were scheduled to be held at the end of the 12-week intervention. (The explanation of the program at the beginning of the intervention was a complex input variable, which had elements of *task input*—the three quality criteria, *group input*—that the work group was told that it would be evaluated as a team, and *environmental input*—that the teams formed two baseball leagues, that scores would be posted, and that there would be rewards for the winning teams. The posted team scores and quality control mea-

Table 2. Variables Investigated in Selected Performance Management Studies

	Stage			
Variable Type	*Input*	*Process*	*Output*	*Feedback*
Task	Task goal[1] Instruction to prepare for the race[2] Instruction to race[2] The group's goal[3] Specification of the three quality criteria[5]	Falls of three or more blocks while the tower was being built[3]	Number of blocks on the tower[3] Load quality[5] Load retention[5] Missort rate[5]	Feedback about task performance[3,4] Posted team scores and quality control measures[5]
Individual	Introduction of behavior standards[1] Telling individual rewards[1] Leader's instruction to an individual crew member[2] Individual goal[3]	Selling, stock work, etc. [1] Verbal behavior of the leader[2]	Intention to continue in the simulation[4]	The actual reward the individual gets[1] Feedback about individual performance[2,3] Positive or negative feedback to the individual[4]
Group	Instructions involving coordination[2] Combination of individual goals[3] Instruction that the work group was a "baseball team"[5]	Cooperative vs. competitive strategy[3]	Alternation of colored blocks on the tower[3]	Feedback about coordination[2] Information about all individuals' performance[3] Distribution of feedback (majority positive or majority negative)[4] Pizza, beer, plaques, and dinner[5]
Environment	Reward policy[1] Instructions that teams form two leagues, that scores would be posted, and that there would be rewards for the winning team[5]	Competition between crews in a boat race[2] Competition between teams in a league[5]	Series standing[2] Coach's rating and ranking of each leader[2]	Posted standings of teams in leagues[5]

Source: [1]Luthans, Paul, & Baker (1985)
[2]Komaki, Desselles, & Bowman (1989)
[3]Mitchell & Silver (1990)
[4]Barr & Conlon (1994)
[5]Kortick & O'Brien (1996)

Table 3. Cells of the Systems Model Investigated in Selected Performance Management Studies

Study	*Input*				*Process*				*Output*				*Feedback*				*Number of cells investigated in each study*
	T	*I*	*G*	*E*	*T*	*I*	*G*	*E*	*T*	*I*	*G*	*E*	*T*	*I*	*G*	*E*	
Luthans et al. (1981)	x	x		x		x								x			5
Komaki et al. (1989)	x	x	x			x		x				x		x	x		8
Mitchell & Silver (1990)	x	x	x		x		x		x		x		x	x	x		10
Barr & Conlon (1994)										x			x	x	x		4
Kortick & O'Brien (1996)	x		x	x				x	x				x		x	x	8
Number of studies that investigated each cell	4	3	3	2	1	2	1	2	2	2	1	1	3	4	4	1	

Notes: T=Task; I=Individual; G=Group; E=Environment.

sures were *task feedback*. The rewards enjoyed by the winning teams—pizza, beer, plaques, and dinner, were *group feedback*, and the posted standings of the teams in the leagues were *environmental feedback*).

Following the introduction of the program, all three quality-control measures increased, although only two of them (load quality and load retention) increased significantly. (This result shows the effect of a combination of several types of *input*, *process*, and *feedback* variables on *task output*). The contributions of instructions, competition, feedback, and tangible rewards cannot be separated in this study, but Kortick and O'Brien (1996) emphasized the competition between teams as a novel aspect of the intervention.

Summary of Variables Investigated in Relation to the 16-Cell Model

Table 2 summarizes the variables investigated in all five studies and classifies them into the 16 cells of the systems model of work group behavior. The four stages of the model are represented in the columns of the table, and the types of variables in each stage are represented in the rows. Footnotes designate the study in which each variable was investigated. The table identifies the variety of variable types that have been manipulated or measured within the scope of performance management. The table also helps to clarify the application of the model to the performance management of teams (1) by providing examples of variables that comprise each cell of the model and (2) by illustrating the relationship between traditional performance management variables (i.e., those focused on tasks and individuals) and the more comprehensive set of variables applicable to groups.

Table 3 identifies the extent to which the various cells of the model were investigated in each study. The table makes clear that Luthans and colleagues (1981) concentrated on individual behavior and individual feedback to the exclusion of group variables. Their independent variables were task inputs and individual feedback. Their only dependent variable was individual selling behavior. On the other hand, Komaki and colleagues (1989) deliberately extended previous work on performance management of individuals to groups by adding measures of group input and group feedback. Mitchell and Silver (1990) carried the investigation of group performance management even further by examining group variables at all four stages. In a more focused experiment, Barr and Conlon (1994) measured the effects of task, individual, and group feedback on individual output. Kortick and O'Brien (1996) were particularly interested in the effect of intergroup competition on group task performance, which accounts for the prevalence of environmental variables in the input, process, and feedback stages, as shown in Table 3.

The row and column totals in Table 3 reveal different aspects of the research on team performance management. The row totals show the completeness of each study in terms of covering the full range of variables relevant to work group behavior. For example, Barr and Conlon (1994) considered only 4 cells while

Mitchell and Silver (1990) addressed 10 cells. On average, about 7 of the 16, a little less than half the cells, were addressed in any one study. The column totals, on the other hand, show how often each type of variable was investigated. The cells most often investigated were task input, individual feedback, and group feedback. This type of input (task) and these types of feedback (individual and group) were the most common types of independent variables in the five studies. In contrast, seven of the cells were relatively neglected; they were each investigated in only one study. The cells least often investigated were primarily in the process and output stages. Typically, these two stages contain the dependent variables in an experiment. Thus, a broad generalization that can be made on the basis of the column totals is that there was more agreement across studies in the types of independent variables that were manipulated than in the types of dependent variables that were measured.

DISCUSSION

Applying the 16-cell model to studies of team performance management has demonstrated the full range of variables that need to be considered when the techniques of performance management are extrapolated from individuals to groups. The ABC model is a powerful tool for modifying individual behavior. Extension of this model to work groups, however, entails manipulating and measuring variables in addition to individual antecedents, behaviors, and consequences. The present analysis has shown that certain independent variables other than individual antecedents and consequences influence behavior, and that certain dependent variables other than individual behavior are influenced by antecedents and consequences.

As shown in Table 3, the selected studies differed in the degree to which they investigated variables beyond the individual level. Luthans and colleagues (1981) focused primarily on individual input, process, and feedback, even though the individual participants were members of a group who worked together in the same department. While the investigators neglected many variables relevant to team performance, they did manipulate task and environmental inputs. In contrast, Komaki and colleagues (1989) made a concerted effort to extend performance management to teams. Their investigation included all four types of variables in the 16-cell model: task, individual, group, and environment.

Mitchell and Silver (1990) found that assigning only individual goals resulted in lower task output (i.e., fewer blocks on the tower) than assigning other types of goals. Task output was higher when the group's goal referred to the overall height of the tower, rather than exclusively to each individual member's contribution. By demonstrating the superiority of task and group goals, Mitchell and Silver showed that focusing solely on individual antecedents and consequences may fail to maximize performance on interdependent tasks.

Barr and Conlon (1994) found an additional limitation of focusing solely on individual feedback, namely, the failure to recognize interaction effects due to group-related variables. Individuals were more likely to continue in the simulation when they received positive feedback about their performance than when they received negative feedback. However, the size of this effect depended on the distribution of feedback within the group. Thus, feedback distribution was a moderator variable because it moderated the effect of individual feedback on behavior. The effect of individual feedback was greater when the majority of the group received positive, rather than negative, feedback. Individual feedback had little effect when the majority of the group received negative feedback. Had Barr and Conlon only measured the impact of individual feedback, they would have missed the importance of the interactions.

Finally, Kortick and O'Brien's (1996) intervention was a complex manipulation, which established quality criteria, teams that competed against one another in "baseball leagues," and rewards for the winning teams. They found that this intervention resulted in a substantial improvement in performance on the criteria, and they concluded that competition was responsible for the effect. However, application of the 16-cell model (as shown in Table 3) reveals that this intervention involved not only an environmental process (competition between teams), but also several different types of input and feedback: task, group, and environment. In order to prove that competition was essential to the intervention, a similar intervention would have to be conducted without the element of competition. Thus, the 16-cell model facilitates the design of alternative experiments to evaluate separately the contributions of different components of complex team-based interventions.

In light of the present analysis, interventions directed at improving the performance of teams should consider all 16 cells of the systems model. Although further research is needed to determine whether comprehensive interventions are more effective than those that manipulate only a few input and feedback variables, studies that measure multiple processes and outputs are clearly more informative than those that measure only a few dependent variables.

CONCLUSIONS

The extension of performance management to teams requires elaboration of the individual-based ABC model, to accommodate the variety of independent and dependent variables that are inherent in organizational work groups. The systems model of work group behavior used in this chapter, was designed to classify and organize the full range of variables that have been identified in the organizational work group literature (Greenbaum & Kaplan, 1997). Its present application demonstrated that every cell of the model is potentially important for team performance management, but that many cells are often neglected in any particular study. In some cases, a performance management intervention focuses on individuals while neglecting the groups to which they belong. Unfortunately, this individual focus may not maximize the performance of the group. In addition, group-related variables can moderate the effects of individual variables. Moderator variables are important because they suppress or magnify the effects of other variables, and failure to recognize their influence can lead to invalid conclusions. Finally, it was suggested that the model could be used to develop more effective interventions by analyzing the components of complex manipulations.

Given its comprehensiveness and its ability to portray the relationships between performance management and work team concepts, the systems model can help to design interventions targeted at improving team performance.

To maximize team performance, interventions should manipulate appropriate antecedents and consequences selected from all applicable input and feedback categories comprising task, individual, group, and environmental variables. Similarly, evaluation strategies should measure the full range of possible outcomes, comprising processes and outputs at the task, individual, group, and environmental levels. Applying the model to the design, implementation, and evaluation of team performance interventions will ensure that they have their intended effects.

REFERENCES

Barr, S.H., & Conlon, E.J. (1994). Effects of distribution of feedback in work groups. *Academy of Management Journal, 37,* 641-655.

Daniels, A.C. (1989). *Performance management.* Tucker, GA: Performance Management Publications.

Daniels, A.C. (1994). *Bringing out the best in people.* New York: McGraw-Hill, Inc.

Deutsch, M. (1949). An experimental study of the effects of cooperation and competition upon group process. *Human Relations, 2,* 199-232.

Druckman, D., Singer, J.E., & Van Cott, H. (1997). *Enhancing organizational performance.* Washington, DC: National Academy Press.

Greenbaum, H.H., & Kaplan, I.T. (1997, August). *A classification system for organizational group variables and their relationships.* Paper presented at the meeting of the Academy of Management, Boston, MA.

Greenbaum, H.H., Kaplan, I.T., & Metlay, W. (1988). Evaluation of problem-solving groups: The case of quality circle programs. *Group and Organization Studies, 13,* 133-147.

Hitchcock, D.E. (1990, September). Performance management for teams: A better way. *The Journal for Quality and Participation,* 52-57.

Johnson, D.W., & Johnson, R.T. (1989). *Cooperation and competition: Theory and research.* Edina, MN: Interaction Book Company.

Johnson, D.W., Maruyama, G., Johnson, R., Nelson, D., & Skon, L. (1981). Effects of cooperative, competitive, and individualistic goal structures on achievement: A meta-analysis. *Psychological Bulletin, 89,* 47-62.

Kaplan, I.T., & Greenbaum, H.H. (1991). A diagnostic model for OD interventions. *Public Administration Quarterly, 14,* 519-532.

Komaki J.L., Desselles, M.L., & Bowman, E.D. (1989). Definitely not a breeze: Extending an operant model of effective supervision to teams. *Journal of Applied Psychology, 74,* 522-529.

Kortick, S.A., & O'Brien, R.M. (1996). The world series of quality control: A case study in the package delivery industry. *Journal of Organizational Behavior Management, 16,* 77-93.

Luthans, F., Paul, R., & Baker, D. (1981). An experimental analysis of the impact of contingent reinforcement on salespersons' performance behavior. *Journal of Applied Psychology, 66,* 314-323.

Metlay, W., Kaplan, I.T., & Rogers, E.E. (1994). Self-management in context. In M.M. Beyerlein & D.A. Johnson (Eds.), *Advances in interdisciplinary studies of work teams: Vol. 1. Theories of self-managing work teams* (pp. 167-185). Greenwich, CT: JAI Press.

Miller, L.K., & Hamblin, R.L. (1963). Interdependence, differential rewarding and productivity. *American Sociological Review, 28,* 768-778.

Mintzberg, H. (1979). *The structuring of organizations: A synthesis of the research.* Englewood Cliffs, NJ: Prentice-Hall.

Mitchell, T.R., & Silver, W.S. (1990). Individual and group goals when workers are interdependent: Effects on task strategies and performance. *Journal of Applied Psychology, 75,* 185-193.

Mohrman, A.M., Jr., Mohrman, S.A., & Lawler, E.E. III. (1992). The performance management of teams. In Bruns, W.J., Jr. (Ed)., *Performance measurement, evaluation, and incentives* (pp. 217-241). Cambridge, MA: Harvard Business School Press.

Skinner, B.F. (1938). *The behavior of organisms.* New York: Appleton-Century-Crofts.

Skinner, B.F. (1953). *Science and human behavior.* New York: Macmillan.

ORGANIZATIONAL CONSIDERATIONS IN THE EVALUATION AND COMPENSATION OF WORK TEAM PERFORMANCE

Duane Windsor

ABSTRACT

This chapter undertakes a conceptual examination of some key issues affecting scholarly and managerial understanding of work team performance evaluation and compensation schemes. Due partly to rapid spread of work teams in U.S. corporations, evaluation and compensation theory and practice in a team context are not yet fully developed. Evaluation and compensation involve comparison of team performance to team objectives, and comparison of team performance to the likely performance of conventional work groups. Reliable evaluation and compensation efforts require a sound theory and operational measures of work team performance including both outcomes and processes. Efficiency and effectiveness issues are involved. Establishment and maintenance of team process is typically more costly and difficult than for conventional work groups. The rationale for work teams is that in the longer run they will perform better under certain conditions.

Advances in Interdisciplinary Studies of Work Teams, Volume 6, pages 139-159.

ISBN: 0-7623-0655-6

INTRODUCTION

This chapter examines the theory and practice of the evaluation and compensation of functioning work teams and their individual members. The chapter does not deal directly with the theoretical desirability or empirical performance superiority of teams, or with their specific internal design for particular work tasks, although those issues are addressed tangentially. Evaluation and compensation determination are post-performance dimensions, but the prospect of evaluation and compensation determination can and should inform specification of work objectives and design of work processes. Evaluation and compensation criteria and practices are partly a function of organizational purpose.

The justification for use of nonhierarchical, empowered team process rather than conventional (i.e., hierarchical) work groups is the expectation that teaming is ultimately (but not necessarily immediately) more efficient and/or effective than the conventional organizational process of supervised individual activity. While there are undoubtedly many circumstances in which teams will outperform groups of individuals, the evidence concerning the performance of nonhierarchical teams relative to conventional (or hierarchical) work groups is more problematic. Team process is typically more costly and difficult in terms of both initial investment and continuing maintenance than conventional work groups. Misapplication of team principles can therefore be wasteful and disruptive (Katzenbach & Smith, 1993). Evaluation of work teams involves comparison of team performance to team objectives, and also comparison of team performance to the likely performance of conventional work groups also achieving a higher degree of accomplishment than barely coordinated individuals. Successful teams presumably overcome certain difficulties inhibiting superior performance by conventional work groups.

One may divide both evaluation and compensation criteria between those concerned with: (1) effort or activity (i.e., process or activity throughputs); and (2) those concerned with accomplishment (i.e., outputs) relative to available critical resources (i.e., inputs). Hence, both effectiveness (the accomplishment of planned goals, objectives, or targets) and relative input-output efficiency are involved in evaluation and compensation judgments.

The chapter proceeds in the following manner after this introduction. The first section makes some basic observations about the business team movement, and the issue of whether there is satisfactory empirical evidence concerning the relative performance superiority of teams. The second section provides definitions of key terms: team, performance, empowerment, and evaluation. The third section provides a conceptual framework of basic ideas and arguments for addressing evaluation and compensation of work team performance. The definition of compensation, reward, or incentive is undertaken here. The fourth section more briefly discusses issues of operationalization and measurement, with particular attention to how the "balanced scorecard" strategic management system of

Kaplan and Norton (1992, 1993, 1996) might be adapted conceptually to the evaluation and compensation of work teams. A brief summary and conclusions section recapitulates the basic findings and arguments of the chapter.

THE TEAM MOVEMENT

The use of nonhierarchical and fully empowered work teams in place of conventional (i.e., hierarchical) work groups has spread rapidly in recent years among major U.S. corporations (Culp, 1995; Mathis & Jackson, 1997). A recent study of 1,800 firms found that about 67 percent use teams in some form; a recent survey of 1,000 factories found that 40 percent of smaller firms used teams, and that 75 percent of firms with over 100 employees used production cell techniques. In 1993, 70 percent of the Fortune 1000 firms reported team-based incentives compared to 11 percent in 1990. Teaming represents a shift from supervised individual activity to empowered group action.

There are two alternative explanations for rapid spread of a profound change in organizational form: (1) demonstrable performance improvements widely adopted (Hendershot & Bailey, 1996); or (2) simple imitation (Drucker, 1995) either in expectation of such improvements, or for other rationales. There has been a strong emphasis on the need for "new order" organizations to meet the challenges of today's "hypercompetitive" business environment (Drucker, 1988; Kanter, 1989; Katzenbach & Smith, 1993; Peters, 1988). Katzenbach and Smith (1993) argue that high-level performance in today's business environment requires key behavioral changes: (1) from individual accountability to individual and joint accountability; (2) from separation of thinking and doing to thinking and doing by everyone; (3) from functional excellence at specific tasks (the hallmark of Taylor's scientific management) to multiple tasks and roles; (4) from managerial control to employee empowerment; and (5) from compensation concerns to personal aspirations and growth. These arguments, even if valid enough in the abstract, mix organizational, group, and individual levels of analysis together, and slide across relatively complex conceptual continua such as that from managed individuals to empowered teams.

It is difficult to gauge the empirical evidence concerning the average (or typical) performance gains resulting from restructuring individual or conventional group work assignments to team approaches. Drucker (1995, p. 97) concluded: "'Team-building' has become a buzzword in American business. The results are not overly impressive." While teams should presumptively outperform individuals in certain situations (Katzenbach & Smith, 1993), it is not strictly clear that teams will necessarily outperform conventional work groups.

One difficulty in interpreting recent experience is that there is a continuum from individual activity through hierarchically coordinated work groups to "true" (i.e., nonhierarchical and fully empowered) work teams. Katzenbach and Smith (1993) at one point use the words *team* and *group* interchangeably. They distinguish pseudo-teams, potential teams, real teams, and high-performance teams in a stage-evolution model.

The notion of stage-evolution in team building is well known in the literature. The models of forming, storming, norming, and performing, or of convening, confronting, collaborating, and completing are fairly familiar to team specialists. The essence of such stage-evolution models is as follows. There is an initial process of assembling a group of individuals that is characterized by forming or convening activities and issues. This assembly is not yet a team. The group will pass through a process of storming or confronting various issues in order to address its assignment. Once the group has developed a set of norms it is engaged in collaboration as distinct from conflict (or storming). At this point, the group has become a team, or least a pseudo-team. A true team will then perform or complete its assignment. Teams require continuing maintenance and renewal efforts. While the group moves in a general sense through these evolutionary stages, there may well be recycling into storming and norming for team maintenance and redevelopment. These stage-evolution models are not meant to imply that all groups ultimately become teams. There can be arrested team development and team destruction (Wetlaufer, 1988, 1994). Empowerment efforts have on occasion failed (Rothstein, 1995).

Team formation may reduce group performance initially. The Katzenbach and Smith model (1993) predicts deterioration of organizational performance (from that achieved by conventional work groups) during the process of team formation. The reason is that forming and storming are costly stages in terms of output performance.

Although there is a relatively large literature on teams (Katzenbach & Smith, 1993), that literature mostly (1) extols the virtues of teams, and (2) provides how-to advice on team building process (Hardaker & Ward, 1987; Katzenbach & Smith, 1993). While Katzenbach and Smith note that "team is a word and concept well known to everyone" (p. 1), they conclude that high-performance teams are in fact extremely rare. This view accords with Drucker's pessimistic evaluation of teams cited earlier. One should distinguish between empirical team performance to date, and the putative promise of future team performance a decade hence when much more may arguably be known about team building. Approaches to measuring team performance are examined in Cohen and Bailey (1997).

Katzenbach and Smith (1993) suggest that demanding performance challenges and a high-performance organizational environment will tend to create effective team process far better than attempting directly to foster a team environment. Ideal teams are, in a crucial sense, those that spring up naturally. Katzenbach and Smith recommend rigorous attention to team basics of "size, purpose, goals, skills, approach, and accountability" as necessary (if not sufficient) conditions for team performance. "Much of the wisdom of teams lies in the disciplined pursuit of performance" (Katzenbach & Smith, 1993, p. 5). They emphasize skill development (at problem solving, technical and functional expertise, and interpersonal relationship building), individual and mutual accountability, and commitment to common goals, approach, and purpose. These conditions are demanding ones.

There are significant examples of organizational changes contrary to the spirit of the team movement. For example, Honda—a well known and highly successful Japanese automobile manufacturing firm—has moved back toward centralization (Chandler & Ingrassia, 1991). Time-and-motion studies on the traditional scientific management (or Taylorism) model have regained some popularity (Adler, 1993). On the other hand, it has been suggested that Toyota, another well known and highly successful Japanese automobile firm, is discovering sociotechnical theory on its own, and so moving to autonomous teams internally (Fujimoto, 1997). Cherns (1976) explicates sociotechnical design theory; Applegate & Cash (1991) provide a good example of sociotechnical theory at work in GE Canada.

Beer, Eisenstat, and Spector (1990) conclude that it is an error to assume simplistically that corporate reorganization is necessary to corporate renewal; reorganization efforts may in fact interfere with the renewal process. These authors argue that the key matter in successful corporate renewal is something other than change in formal organizational structure. Managers often assume that, where changing markets and competition dictate organizational change, the first step is formal restructuring on the following expectation (where the symbol $X \rightarrow Y$ denotes that X logically implies and realistically influences X and the symbol Δ denotes "change in" X):

$$\begin{gathered} \text{structure} \rightarrow \text{behavior} \rightarrow \text{performance} \\ \Delta\,\text{structure} \rightarrow \Delta\,\text{behavior} \rightarrow \Delta\,\text{performance} \end{gathered} \tag{1}$$

The expectation is not necessarily false, but neither is it necessarily true. The matter is conditional upon circumstances. Two well known but very different corporate examples may be noted here: (1) Motorola (Gogan, 1994a; Gogan, 1994b); and (2) General Electric (Applegate & Cash, 1991). Current examples from even the best companies may not validly predict what success will require a decade hence.

Motorola's Changes

Motorola received the 1988 Malcolm Baldrige National Quality Award (Therrien, 1989). A decade earlier, in 1979, at a meeting of Motorola executives, a sales manager argued: "Our quality stinks! That's our problem, and if we don't do something about it the Japanese will continue to take market share away from us" (Gogan, 1994a, pp. 1-2). Motorola, under Robert (Bob) Galvin (CEO, 1959-90), son of the firm's founder Paul Galvin (CEO, 1928-59) undertook to improve product quality tenfold and to upgrade employees' skills within five years. In 1987, Motorola announced three additional successive two-year tenfold increases in product quality through 1992 to move toward a zero defect target. Every employee was required to take a new quality course. This effort was conjoined with a planned reduction of product-to-market cycle time by tenfold within five years.

In 1980, Galvin established a Motorola Training and Education Center (MTEC) that subsequently (1989) became Motorola University. In 1983, Galvin began restructuring Motorola into smaller, focused business units with greater authority and undertook delayering to remove two organizational levels. In 1984 every group was required to spend at least 1.5 percent of payroll on training (Gogan, 1994a).

The firm already had a participative management program (PMP), established in the late 1970s, in which PMP "teams" (i.e., Japanese-style "quality circles" working on common "line of sight" product or activity) met weekly to review performance. A 200-page manual covered procedures, and monthly PMP bonuses in the U.S. (but not abroad) attempted to link team goals and compensation. PMP courses supported team operation dimensions. There were employee complaints about the allegedly mechanistic format of the PMP approach and the unfairness of the PMP bonus plan. This PMP compensation scheme was replaced in 1987 by a return on net assets (RONA) bonus scheme that specified targets for both the corporation and the business unit. The scheme attempts to link individual and team performance to organizational results measured in terms of financial outcome relative to resources deployed, without respect to individual and group effort.

In 1989, a total customer satisfaction (TCS) team program replaced PMP. By 1993, Motorola had 4,000 temporary TCS teams, and team projects had reportedly saved over $2 billion in manufacturing costs since 1989 (Gogan, 1994a). A corporate-wide TCS team competition became an annual event. The effort was considerably more advanced among product design and manufacturing teams than among administrative teams.

The next phase became an effort at work team empowerment (i.e., a shift to self-managed teams through delayering of supervisory levels). TCS teams had previously required managerial approval of recommendations. Empowerment became a key initiative. A central issue became whether the approach of procedures, goals, and metrics—which had been used in PMP, quality improvement, and cycle time improvement efforts—could be used in empowerment. Katzenbach and Smith (1993) report on the Connectors Team within the Government Electronics Group.

Motorola's revenues rose from nearly $5.9 billion in 1986 to an estimated $30 billion in 1996. Employees rose from 105,000 in 1990 to 120,000 in 1993; net sales were $10.9 billion (about $103,800 per employee by calculation) in 1990 and $17 billion (about $141,700 per employee by calculation) in 1993 (Gogan, 1994b). Explosive growth occurred during 1992-1995 in cellular phones and pagers (Takahashi, 1996). Recently, however, growth has tapered off, and competition has intensified in the cellular phone business (Hardy, 1996); Motorola's "lead has since eroded, price wars are now endemic and the company has been slow to find new markets or deliver breakthrough products in the old ones" while memory-chip prices fell 75 percent in a year (Takahashi, 1996). Slow government deregulation and slow cable network improvements have been suggested as important factors (Takahashi, 1996). Christopher Galvin, then Chief Operating Officer (COO), became Chief Executive Officer (CEO) at the beginning of 1997. The future success of Motorola arguably depends on the introduction of new technologies (Hardy, 1997).

General Electric's Changes

In the 1980s, General Electric (GE), under a new CEO John (Jack) Welch, undertook a drastic program of head-count downsizing, organizational redesign, and business process reengineering. The general objective was to create a true learning organization (Garvin, 1993). Welch stated (1988) that, given a slower growing and intensively competitive global economy, "We had to find a way to combine the power, resources and reach of a big company with the hunger, the agility, the spirit and the fire of a small one" (Applegate & Cash, 1991, p. 1). Welch set goals of being number one or two in each market. Corporate restructuring divested $10 billion in assets and made $20 billion in acquisitions (Katzenbach & Smith, 1993). Corporate head count was reported as 300,000 in 1989 and 239,000 in 1997 (roughly a 20 percent decline).

Within Pooled Financial Services (PFS) of GE Canada, located at Toronto, two organization levels and some 40 percent of employees were eliminated, while productivity and work quality rose (Applegate & Cash, 1991). PFS was created in 1985 through centralization of financial, administrative, and information technology services. PFS head count fell from 360 (March 1985) to 218 (March 1987) six months ahead of forecast. This pooled organization was restructured into five centers (control, analysis, employee, facilities and support, and information) together with human resources support. Each center was restructured into self-managed work teams (a total of 12 outside of human resources support). The PFS was directed by a coordinators' council of a manager, five coordinators, and two human resources support personnel. Teams developed mission statements, goals and objectives, work plans, budgets, and job designs. Team process became linked to a team and individual performance appraisal system and a supportive reward system.

During the years 1987-1996, GE revenues (including financial services) increased from $39.3 billion to $79.2 billion (split roughly in the latter year between $46 billion from sales and $33 billion from financial services). Net income rose from $2.33 billion (5.9% of revenue by calculation) to $7.28 billion (9.2% of revenue by calculation). Earnings per share (EPS) rose from $2.119 to $4.40; however, GE engaged in a substantial share repurchase program so that the denominator for EPS declined over this period. It is estimated that GE will have achieved $7-10 billion in cumulative savings by 2002 through the combination of head-count downsizing and efficiency improvements. There is clearly a strong incentive to pass greater revenue volume through fewer employees, so as to increase revenues per employee. The ratio for 1997 is about $331,381 overall by calculation, and perhaps some $192,470 by calculation for sales alone (excluding financial services)—and hence perhaps stronger than at Motorola. The calculation does not remove financial services personnel.

The Motorola and GE examples certainly illustrate the principle that organizational transformation efforts can lead to substantial revenue and profit gains. In both companies, new initiatives undertaken by the chief executive officer paid off dramatically. It is, however, clearly difficult to isolate the contributions made by each of the specific dimensions involved in the transformation effort. At Motorola, downsizing was not pursued; on the contrary, there was substantial investment in personnel capabilities. One may speculate that Motorola's success was due to external forces that greatly increased demand for certain products, although its share of that demand may have depended on internal quality improvements. Teams appear to have saved some money for the firm, but there is little evidence that team structure per se could stave off the effects of the slowdown in industry demand. At GE, downsizing was a key dimension of organizational transformation, and team structure appears to have been a consequence of that dimension. While GE has reported dramatic cost savings, it is difficult to disentangle the downsizing and efficiency dimensions. The two examples suggest that how one defines key notions such as team, empowerment, and performance is a critical matter in testing the validity of team theories. The next section addresses the definition of key terms.

DEFINITIONS OF KEY TERMS

Three notions require careful attention to definition and interpretation: what is a team; what is empowerment; how does one evaluate performance? Each of these terms is defined in turn below.

Katzenbach and Smith (1993, p. 45) provide the following succinct definition of a *team*: "A team is a small number of people with complementary skills who are committed to a common purpose, performance goals, and approach for which they hold themselves mutually accountable." A *team* is then a small and cross-functional group of people organized to work together in a different way relative to a conventional work group. (In this sense, a large corporation as a complex organization cannot be a "team.") This different way of working turns on shared leadership and shared commitment within a cross-functional group. Hence team members are specialists, but their work is coordinated within the group rather than by hierarchy. In a sense, the team receives a mission (with objectives) and figures out how to accomplish the mission. A team is a directional (i.e., purposive) rather than a directed (i.e., supervised) work group.

The distinction between a "true" team (whether functioning face-to-face in close proximity or as a virtual reality through rapid communication channels) and a conventional work unit turns on notions of: (1) shared purpose, goals, and approach; (2) complementarity of skills; and (3) personal acceptance of joint responsibility. A group is simply a set of individuals without appropriate commonalities or complementarities, and presumably without individual responsibility acceptance. In a true team environment, there are connotations of increased collaboration, greater self-management (a nonhierarchical work unit), and greater learning capacity. The team notion embeds expectations that conventional (or hierarchical) work units involve greater interpersonal and intergroup conflict, as well as slower individual and group learning (cf. Katzenbach & Smith, 1993). The argument is that a high-performance team will collaborate and learn more effectively than a hierarchical work group. It is not transparent from the learning literature that this argument must be strictly valid.

These elements form what can be understood as a continuum of organizational forms from a traditional enterprise structure (one characterized by extreme vertical hierarchy, high if suppressed internal conflict, and slow learning speed) to an ideal, or perhaps idealized, team environment (one characterized by formal equality, high collaboration accelerated by directive internal conflict, and rapid learning speed). Forms between these two extreme examples involve various combinations of hierarchy, collaboration, and learning capacity. So, teams may be managed hierarchically in an intermediate example, rather than being self-directed. It is not surprising then that Mohrman, Cohen, and Mohrman (1995) suggest that hierarchy can embed or be composed of teams. A large and hierarchical organization may not be, strictly speaking, a "team" of teams, but the organization can be composed of many teams.

The proposed continuum of organizational work forms is illustrated in Figure 1. The horizontal axis is a continuum between strictly hierarchical organization and strictly self-managed work teams. This axis expresses the essence of the team structure notion. The vertical axis, which addresses the essence of the team process notion, is more difficult to explain. The sense of the vertical axis is that there is a continuum between destructive conflict and constructive collaboration. Neither conflict nor collaboration is inherently good or bad. There may not be, strictly speaking, a continuum between conflict and collaboration. The team process is rather one in which collaboration arises out of conflict. The conflict-to-collaboration evolution may fail. Teams may be destroyed by internal conflict or lack of external support. Team theory stresses the positive aspects of conflict in team settings and the negative aspects of conflict in non-team groups. As Manz and Neck (1995) and Manz and Stewart (1997) point out, "teamthink" may contain the seeds of "groupthink." Self-managed work teams are particularly prone to the groupthink, or hasty consensus, syndrome. The object of teamthink, properly

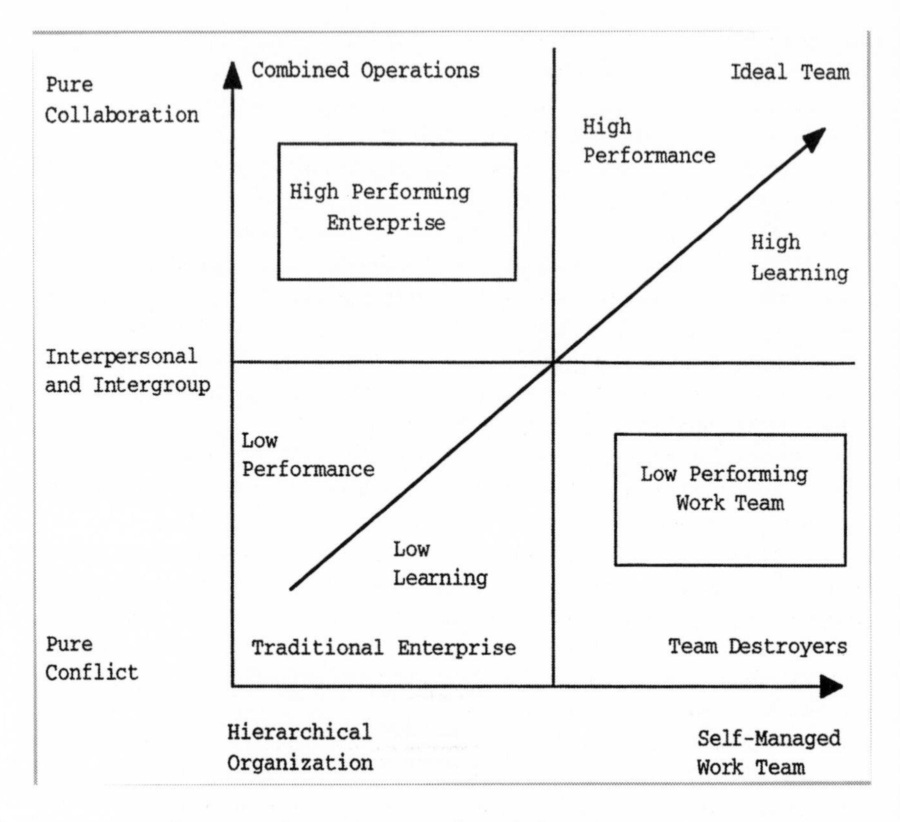

Figure 1. A Continuum of Work Organization Forms

evolved, is constructive thought patterns. Conventional work units may collaborate successfully together in temporary combined operations without being true teams. The "team" approach, therefore, is an effort to improve group collaboration and learning through the positive use of conflict. In light of these considerations, Figure 1 emphasizes the distinction between constructive forms of collaboration and destructive forms of conflict.

Empowerment (Frey, 1993) appears to be an inherent and defining characteristic of work teams by way of distinction from conventional work groups. There are, however, at least two difficulties with the notion of empowerment. One difficulty is that there are at least two radically different frameworks for defining empowerment. Another difficulty is that there appear to be levels of empowerment within an organization.

Different authors have taken two radically different approaches to defining empowerment (cf. Fisher, 1997). One framework emphasizes delegation of authority to act (see Maremont's [1995] account of CEO George Fisher's empowerment effort at Kodak). Another framework emphasizes enhancing employees' capabilities to make a difference in goal attainment (see Mohrman et al., 1995, chap. 9). The distinction corresponds to the common sense (or dictionary) difference between empowerment as giving power or authority to another, and empowerment as enabling another. Embedded within the distinction is the difficulty of defining *power* in the first instance. Power has one connotation of control over others, so that empowerment is the abdication of such control. Power has another connotation of ability to control events, so that empowerment is the creation of the ability in others to affect events. This distinction is not resolved in the empowerment literature.

Both notions of empowerment were involved at Motorola. In addition to educational efforts and team structure, Motorola identified some seven levels inherent in its situation (Gogan, 1994a): (1) supervisor assignment of work; (2) operator determination of training needs; (3) supervisor request for input; (4) team ownership for performance; (5) team interfacing with other actors; (6) team decision making; (7) fully autonomous teams using external support functions as consultants. The then CEO Gary Tooker observed that empowerment by decree was an "oxymoron," since empowerment had to emerge by employee acceptance of responsibility. (Gogan, 1994a)

Stated in very general terms only, there are two quite different types of teams: (1) interchangeable members; and (2) functionally specialized members. An instance of the first type is an accounting unit, all of whose members can handle all accounting matters coming to the unit. A tennis doubles pair is this type of team; each player must serve and receive. Skill level may well vary, but work is rotated rather than specialized. Specialization occurs where sufficient variation in skill level is evident. An instance of the second type is a true cross-functional team. A baseball or football team is this type; each player occupies a specialized position. Drucker (1995) specifies three types of teams: (1) tennis doubles; (2) baseball; and (3) football. Within the cross-functional team approach, baseball is played one specialist at a time, while football is played conjointly on signal. There is reasonably clear hierarchy of planning (i.e., ex ante decision making) in football (e.g., the quarterback on offense) with specialized squads for offense (and a specialized kicking squad within offense) and defense. Despite the manager's strategic role in baseball, the individual player acts alone according to circumstances rather than to plan, and plays both offense and defense until relieved. Tennis doubles teams rotate roles on offense and defense.

Katzenbach and Smith (1993) distinguish between teams that make recommendations (e.g., task forces, project groups, audit/quality/safety groups) and teams that make or do things of direct value-added impact (e.g., research and development, operations, marketing, service). Sundstrom, DeMeuse, and Futrell (1990) describe four kinds of teams: advice and involvement (e.g., quality control circles and committees); production and services (e.g., assembly groups and sales teams); projects and development (e.g., engineering and research groups); action and negotiation (e.g., sports teams and combat units). Cohen and Bailey (1997) distinguish work teams, parallel teams, project teams, and top management teams.

The term *performance* takes the general meaning of functional effectiveness in accomplishing some identifiable task through execution of an action or a process. Definition, measurement, and evaluation of performance involve at least the following eight identifiable elements or components:

1. The particular *task* (whether broadly or narrowly defined) to be accomplished;
2. The outcome, target, goal, or *objective* (whether quantifiable or qualitative) specified in connection with the task;
3. The action or *process* executed to achieve the target or outcome;
4. The *time* horizon over which the action or process is executed and the outcome is measured and evaluated;
5. The *resources* available for execution of the task;
6. The non-resource *constraints* on execution;
7. Input-output or *technical efficiency* in resource usage;
8. Benefit-cost or *economic efficiency* in resource usage.

In general, definition involves the identification and specification of task, outcome, action, process, time horizon, resources, constraints, and efficiencies. Those elements susceptible to cardinal or at least ordinal measurement must be operationalized and metricized.

Evaluation may take two approaches. One approach is to focus on outcomes in relationship to efficiency of resource utilization. This approach can be either a single outcome or a multiple outcomes notion. For example, return on investment or ROI (a per dollar measure) is a narrow financial outcome conception that embeds an implicit economic efficiency measurement. The return is not simply total profit, but rather the profit per dollar of investment. One can expand the single outcomes notion to some more complex construct. Such an approach would link a number of performance outcome measures that ideally point in the same direction under optimal circumstances. The various outcomes measures may be commensurable (in which case a weighted performance index can be constructed), or they may be, as in the Kaplan and Norton "balanced scorecard" (1992, 1993, 1996) approach discussed below, incommensurable (in which case a qualitative judgment concerning relative importance must be made). The other approach requires an even more complex construct comprised of both process and outcomes measures. The Kaplan and Norton balanced scorecard is constructed in this manner, and the various dimensions in the proposed measurement system are necessarily incommensurable—although they may still point in the same direction under certain circumstances.

TEAM THEORY: A CONCEPTUAL FRAMEWORK

This chapter addresses team theory from a relatively simple perspective concerning strategic management of the firm for the creation and maintenance of sustainable competitive advantage. This perspective involves formulating a causal theory of organizational advantage and the role of individual behavior within organizational advantage. The perspective is laid out initially without reference to the role of team structure and process. Two brief examples will serve here: the "simple scheme" and the Wal-Mart model.

The "simple scheme" (Peters & Austin, 1985) portrays organizational performance as the outcome of an interaction among four key factors: (1) constant product (and/or service) and process innovation; (2) customer satisfaction; (3) employee motivation for and skill at performance of innovation and customer satisfaction activities; and (4) appropriate management style identified as "management-by-wandering-around" (MBWA) in the Peters and Waterman (1982) approach to corporate excellence. This "simple scheme" is depicted by Peters and Austin as a circle (for management style) within a triangle of the other three factors defined so that employees are the "base" (and key resource) of the enterprise.

Sam Walton described his perception of the keys to the success of Wal-Mart as follows: (1) employee motivation through empowerment and compensation; (2) technological superiority (information technology for inventory management and tracking "traits" affecting individual store profitability); (3) pricing flexibility at store level; (4) stringent cost control and careful supplier negotiation; (5) hands-on management (MBWA); and (5) the resulting stakeholder (customer, employee, and supplier) loyalty to Wal-Mart as preferred company (Foley & Mahmood, 1994).

Neither of these theories of organizational advantage and individual behavior addresses the role of teams. Team theory is about individual performance incentives within organizational architecture. (The term *architecture* here has a broader meaning than simply formal structure). The team constitutes one type of linkage between individual and organizational architecture. The broader theory on which team evaluation and compensation determination must draw is therefore of the general form below:

$$\text{context} \rightarrow \text{tasks} \rightarrow \text{structure} \rightarrow \text{behavior} \rightarrow \text{performance} \tag{2}$$

It also follows that:

$$\begin{aligned} &\Delta\ \text{context} \rightarrow \Delta\ \text{tasks} \rightarrow \Delta\ \text{structure} \\ &\rightarrow \Delta\ \text{behavior} \rightarrow \Delta\ \text{performance} \end{aligned} \tag{3}$$

One can say further that, in reverse order of logic and influence, performance (the final dependent variable) is a function of behavior, which is, in turn, a func-

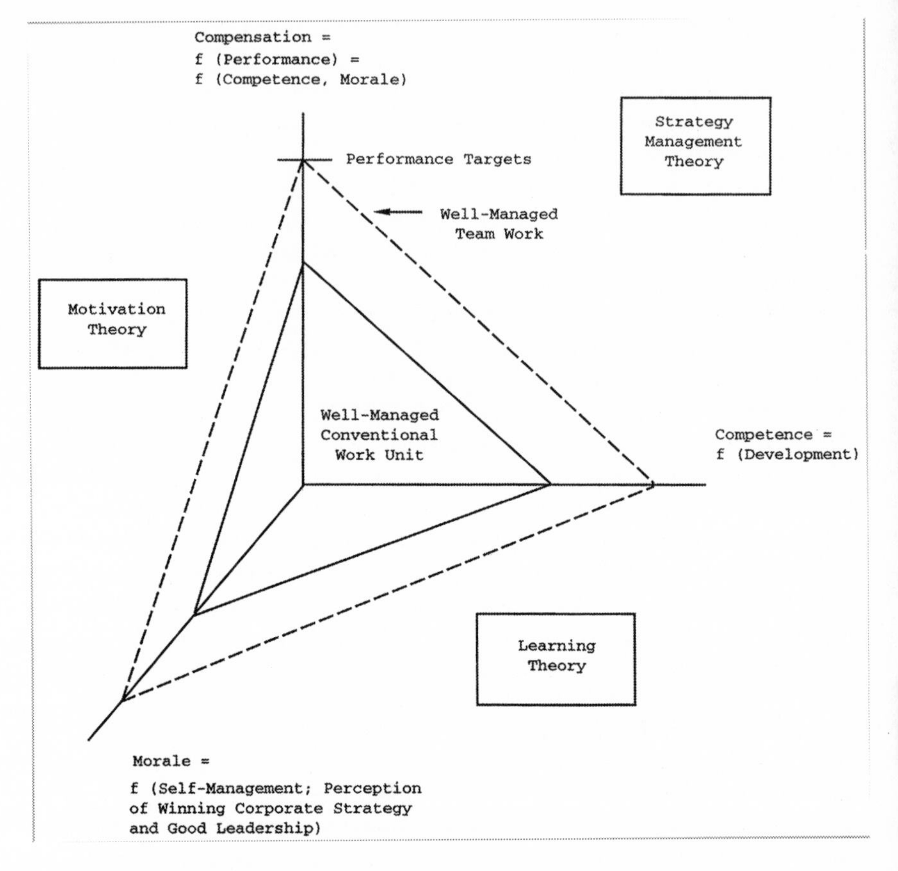

Figure 2. Key Arguments of Team Theory

tion of structure, and so on back to context. Behavior (or conduct) is also affected by morale and compensation, not shown directly above but discussed in detail later.

Context is a rubric notion embodying the interaction between environment and organization that logically and realistically generates the firm's strategy. That strategy implies some definable set of employee tasks to be structured in some form.

Key Arguments of Team Theory

Team theory asserts that a particular organizational structure, process, and culture (at least for certain contexts) will enhance behavior and therefore performance. Key arguments of team theory are illustrated in Figure 2.

This figure provides a conceptual model for integrating learning theory, motivation theory, and strategic management theory. The figure comprises three axes labeled compensation, competence, and morale. Learning theory links competence development and morale. Motivation theory links morale with compensation (and hence with performance). Strategic management theory links competence and performance. Effective teams presumably exhibit higher competence, higher morale, and higher performance than conventional work groups. Performance targets are stipulated on the compensation axis for ease of presentation in three dimensions.

The essential claim is that a well-managed work team will outperform a well-managed conventional work unit. (There is an additional claim that conventional work units tend to be less functional in reality). Otherwise, teams are superior to barely coordinated individuals engaged in separate tasks. Obviously, poorly prepared or internally conflicted teams may underperform well-managed conventional work units.

The model stipulates seven variables: goal (performance is evaluated relative to goal), performance, motivation, competence, morale, compensation, and development. Goal becomes subsumed in the notion of performance targets. (The performance targets move toward the stipulated goal or goals). This model is shown below in functional form (where X is a function "f" of Y):

$$\begin{aligned} \text{performance} &= f\,(\text{motivation, competence}) \\ \text{motivation} &= f\,(\text{morale, compensation}) \\ \text{competence} &= f\,(\text{development}) \end{aligned} \tag{4}$$

It requires four arguments to integrate learning theory, motivation theory, and strategic management theory. The key arguments are: (1) team competence is a function of personnel selection and development; (2) team morale is a function of self-management and perception of winning corporate strategy and good leadership (these latter two factors may occur in the instance of conventional work units); (3) team performance is a function of competence and morale; and (4) individual compensation should be a function of team and corporate performance. (The difficulty in defining compensation is that it depends both on effort and performance outcomes; the latter acquires the resources to reward the former).

Evaluation and compensation schemes have typically been geared to individuals rather than to teams (Katzenbach & Smith, 1993). *Compensation* (i.e., reward or incentive) becomes more difficult to define in a team context. The greatest reward for team members may be the team itself—its shared activity and performance success (Katzenbach & Smith, 1993). A particularly difficult problem in team performance evaluation concerns promotion of team members (Mahoney, 1996). Team process emphasizes harmony and collaboration in group-directed activities. How does top management then evaluate individuals for promotion to various forms of supervisory responsibilities?

One may think of organizational performance as a partial function of employee performance. (Organizational performance is affected by a variety of other independent variables). Motivation theory concerns the linkages among employee morale, employee performance, and employee compensation. The relationship is not simply linear in the sense that employee performance is a function of employee morale which is, in turn, a function of employee compensation as illustrated schematically in the following model:

$$\begin{gathered} \text{compensation} \rightarrow \text{morale} \rightarrow \text{performance} \\ \Delta\,\text{compensation} \rightarrow \Delta\,\text{morale} \rightarrow \Delta\,\text{performance} \\ \text{performance} = \text{f (morale)} = \text{f (compensation)} \end{gathered} \tag{5}$$

It is more likely that the three factors interact simultaneously as follows:

$$\text{performance} = \text{f (morale, compensation)} \tag{6}$$

Performance is a complex notion, since goal clarity, efficacy, resources, and competency are also involved. The purpose here is not to give an account of organizational performance, but rather of the relationship of performance to morale and compensation. Performance occurs at the individual, group or team, and organizational levels. Compensation is a partial function of organizational performance, which in turn is a partial function of employee performance. Employee performance affects morale, as well as morale affecting performance. A schematic illustration would thus involve a triangular display with dual-directional arrows running between each of the three factors.

While the relationship between high morale and high performance is arguably well established, the relationship between compensation and morale (and thus between compensation and performance) is a more difficult matter. A body of literature challenges the conventional assumption that either higher compensation, or changes in format of compensation, result in permanent increases in employee and thus organizational performance.

The conventional performance appraisal process has come under severe criticism (Bruns, 1992; Ilgen, Barnes-Farrell, & McKellin, 1993; Levinson, 1976; McGregor, 1972). At Nordstrom, a very successful department store chain, there was in 1989 a widespread assault on the company's performance measurement system by employees, unions, and government regulators (Simons & Weston, 1991). The key element of the system was "sales-per-hour" monitoring and compensation.

The essential purpose of any appraisal system is to focus managerial attention on the relationship between employee behavior and performance results as a basis for promotion and compensation decisions. Thompson and Dalton (1970) distinguish between "zero-sum" and "management by objectives" (MBO) appraisal systems.

A zero-sum system is one in which each employee is ranked against all other employees. (One may think of an index that sums to, say, 100 points for all employees. If one employee receives a rating of 5, then only 95 points remain to be distributed among all the other employees). All employees are necessarily forced to compete with one another, with potentially very negative motivational consequences destructive of organizational performance objectives.

A management-by-objectives system emphasizes establishment of specific goals for each employee against which performance is then measured. MBO systems have their own problems, of course, in terms of goal setting (employees want to exceed relatively low goals, and employers want to stretch employees toward higher goals) and performance definition and measurement. But in Thompson and Dalton's argument, an MBO system can be a form of "positive-sum" game in which the overall index can in effect exceed 100 (and desirably so). All employees can potentially receive a rating of 5 through personal growth and improvement.

Recently, Kohn (1993) has argued against conventional wisdom that the promise of rewards does not in fact commonly function as a performance incentive. He summarizes evidence from laboratory and workplace studies as suggesting that incentive systems tend to undermine performance enhancement efforts. Incentive systems, in his view, achieve at best purely temporary compliance with managers' wishes. Incentive systems tend to punish rather than to motivate employees (because incentives are largely threat of loss), to destroy rather than to nurture interpersonal relationships (through creation of zero-sum situations), and to discourage employee risktaking and interest (because commission failure is punished more reliably than omission failure). Kohn (1993) regards incentives as barely disguised bribes. Responses by other specialists may be found in "Rethinking Rewards" (Stewart et al., 1993).

OPERATIONALIZATION AND MEASUREMENT

The interaction among compensation, competence, morale, and performance is possibly of the general form shown in Figure 3. The depiction shows a simultaneous interdependence of each factor ultimately affecting the other three factors. The conception is similar to that of Kaplan and Norton's (1992, 1993, 1996) "balanced scorecard" system of metrics.

Team evaluation may well require something akin to Kaplan and Norton's approach. As originally designed for purposes of strategic management, the balanced scorecard combines measures of financial performance (i.e., bottom-line outcome), customer satisfaction (i.e., the factor that underlies the firm's revenue flow and margin), organizational excellence (i.e., roughly expense control and time-to-market control), and economic value creation (i.e., product-process innovation and learning outcomes). Each component of the scorecard measures per-

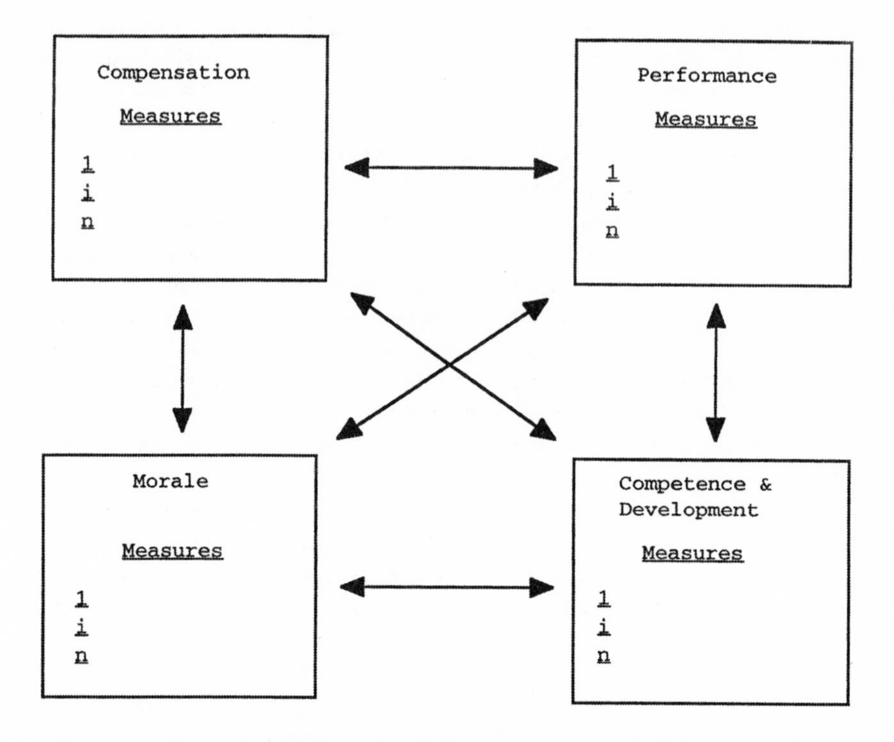

Figure 3. A "Balanced Scorecard" Measurement of Team Performance

formance outcomes against goals. The four principal components are interdependent. Kaplan and Norton (1992) portray four separate scorecards (each one internally linking specific goals and measures), mutually linked (through six dual-directional arrowheads in total) into a system of metrics. They characterize these four components as perspectives (which this author reinterprets slightly here) taken by four different types of participants in the firm: (1) shareholders (financial perspective); (2) customers; (3) employees (excellence); and (4) management (value creation through innovation and learning).

A key feature of the approach is avoidance of information overload through selection of vital measures. Kaplan and Norton (1992) think of the balanced scorecard as akin to the dials and indicators of an aircraft cockpit providing information on many aspects of a flight. These dials and indicators provide summary information on "the current and predicted environment" of the flight (Kaplan & Norton, 1992, p. 72). Reports on fuel, speed, altitude, and so on are not commensurable and hence are not reducible to a single index. As a result, "Reliance on [any] one instrument can be fatal" (p. 72). Complexity of management of a task (or an organization) implies complexity of performance definition, measurement, and evaluation. Following the same analogy, team operation involves multi-person crews interacting with ground personnel for successful flight operations in place of substituting several single-person flights not interacting with ground personnel.

Meyer (1994) applies what is essentially a balanced scorecard concept (i.e., "dashboard" gauges) to team performance. He points out that what matters in team empowerment is information, not simply about results (which are important to the firm's top management), but also about team process: "senior managers should create the strategic context [i.e., strategic goals] for the teams but not the measures" (Meyer, 1994, p. 101). A dashboard is therefore both team specific and team designed (within stipulated strategic goals). He reports a dashboard for a product development (disk drive) team at Quantum Corporation (Milpitas, California) that would keep track of: (1) in the center of the dashboard—product and process development status by time (critical path to completion), weekly hot item(s), and status of evaluations in process; (2) employee satisfaction and recent change in satisfaction, and staffing (relative to goal); (3) time period to next business review with higher management; (4) product costs (materials and overhead) and program costs to date (relative to plan); (5) product quality; and (6) business margin and revenue (current relative to plan) (Meyer, 1994). Meyer (1994, p. 100) defines four basic steps for process measurement: (1) identifying the factors important to customer satisfaction (i.e., "time, cost, quality, and product performance"); (2) mapping the cross-functional process for delivering those factors; (3) identifying "the critical tasks and capabilities" for process completion; and (4) creating measures for tracking those tasks and capabilities.

SUMMARY AND CONCLUSIONS

This chapter has examined conceptually some presumptively important issues in the evaluation and compensation of work teams. The team movement has spread rapidly in business, faster than the development of relevant evaluation and compensation theory and practice. The chapter suggests that the empirical evidence favoring teams over conventional work groups is not yet well established, although teams often clearly outperform individuals in certain circumstances. Head-count downsizing often forces team structure design in order to move more volume through fewer employees. It is desirable to develop a stronger theoretical framework for team design and operation as a basis for the design of evaluation and compensation schemes. Recent scholarship has questioned traditional assumptions about compensation and morale, and also the linkages among corporate performance, organizational structure, and individual behavior. The chapter suggests that an analog to the Kaplan and Norton (1992, 1993, 1996) balanced scorecard approach developed for strategic management may prove useful to future research and reflection. Meyer (1994) reasonably illustrates the analog.

REFERENCES

Adler, P.S. (1993). Time-and-motion regained. *Harvard Business Review, 71*(1), 97–108.

Applegate, L.M., & Cash, J.I. (1991). *GE Canada: Designing a new organization.* Harvard Business School Case 9–189–138.

Beer, M., Eisenstat, R.A., & Spector, B. (1990). *The critical path to corporate renewal.* Boston: Harvard Business School Press.

Bruns, W.J. (Ed.). (1992). *Performance measurement, evaluation, and incentives.* Boston: Harvard Business School Press.

Chandler, C., & Ingrassia, P. (1991, April 11). Shifting gears: Just as U.S. firms try Japanese management, Honda is centralizing. *Wall Street Journal,* A1.

Cherns, A. (1976). The principles of sociotechnical design. *Human Relations, 78*(8), 783-792.

Cohen, S.G., & Bailey, D.E. (1997). What makes teams work: Group effectiveness research from the shop floor to the executive suite. *Journal of Management, 23*(3), 239-290.

Culp, M.L. (1995, December 24). Use teams to build an advantage. *Houston Chronicle,* 1EE.

Drucker, P.F. (1988). Coming of the new organization. *Harvard Business Review, 66*(1), 45-53.

Drucker, P.F. (1995). There's three kinds of teams. In P.F. Drucker (Ed.), *Managing in a time of great change* (pp. 71-102). New York: Truman Talley Books/Dutton.

Fisher, J.R. (1997). A culture of contribution. *Executive Excellence, 14*(1), 16.

Foley, S., & Mahmood, T. (1994). *Wal-Mart Stores, Inc.* Harvard Business School Case No. 9-794-024.

Frey, R. (1993). Empowerment or else. *Harvard Business Review, 71*(5), 80-94.

Fujimoto, T. (1997). *Transforming automobile assembly: Experience in automation and work organization.* New York: Spring-Verlag.

Garvin, D. (1993). Building a learning organization. *Harvard Business Review, 71*(4), 78–91.

Gogan, J.L. (1994a). *Motorola: Institutionalizing corporate initiatives.* Harvard Business School Case 9-494-139.

Gogan, J.L. (1994b). *Motorola Corporation: The view from the CEO office.* Harvard Business School Case 9-494-140.

Hardaker, M., & Ward, B.K. (1987). How to make a team work. *Harvard Business Review, 65*(6), 112-117.

Hardy, Q. (1996, November 15). Motorola selects Christopher Galvin, grandson of firm's founder, as CEO. *Wall Street Journal,* A3.

Hardy, Q. (1997, February 25). Motorola is losing sales in early round of digital phone market, report says. *Wall Street Journal,* B6.

Hendershot, A., & Bailey, G. (1996, August 26). How we brought teamwork to marketing. *Wall Street Journal,* A10.

Ilgen, D.R., Barnes-Farrell, J.L, & McKellin, D.B. (1993). Performance appraisal process research in the 1980s: What has it contributed to appraisal in use? *Organizational Behavior and Human Decision Processes, 54*(3), 321-368.

Kanter, R.M. (1989). The new managerial work. *Harvard Business Review, 67*(6), 85–92.

Kaplan, R.S., & Norton, D.P. (1992). The balanced scorecard: Measures that drive performance. *Harvard Business Review, 70*(1), 71-79.

Kaplan, R.S., & Norton, D.P. (1993). Putting the balanced scorecard to work. *Harvard Business Review, 71*(5), 134-142.

Kaplan, R.S., & Norton, D.P. (1996). Using the balanced scorecard as a strategic management system. *Harvard Business Review, 74*(1), 75-85.

Katzenbach, J.R., & Smith, D.K. (1993). *The wisdom of teams: Creating the high-performance organization.* Boston: Harvard Business School Press.

Kohn, A. (1993). Why incentive plans cannot work. *Harvard Business Review, 71*(5), 54-63.

Levinson, H. (1976). Appraisal of what performance? *Harvard Business Review, 54*(4), 30-36, 40, 44, 46, 160.

Mahoney, R.J. (1996). *Beyond empowerment: Relevant ad hockery* (CEO Series Issue No. 1). St. Louis, MO: Washington University, Center for the Study of American Business.

Manz, C.C, & Neck, C.P. (1995). Teamthink: Beyond the groupthink syndrome in self-managing work teams. *Journal of Managerial Psychology, 10*(1), 7-15.

Manz, C.C., & Stewart, G.L. (1997). Attaining flexible stability by integrating total quality management and socio-technical systems theory. *Organization Theory, 8*(1), 59-70.

Maremont, M. (1995, January 30). Kodak's new forces. *Business Week,* 62-68.

Mathis, R.L., & Jackson, J.H. (1997). *Human resource development.* Minneapolis, MN: West Publishing Co.

McGregor, D. (1972). An uneasy look at performance appraisal. *Harvard Business Review, 50*(5), 133-138 (original work published 1957).

Meyer, C. (1994). How the right measures help teams excel. *Harvard Business Review, 72*(3), 95-103.

Mohrman, S.A., Cohen, S.G., & Mohrman, A.M., Jr. (1995). *Designing team-based organizations: New forms for knowledge work.* San Francisco: Jossey-Bass.

Peters, T. (1988). Restoring American competitiveness: Looking for new models of organizations. *Academy of Management Executive, 2*(2), 103-110.

Peters, T., & Austin, N.K. (1985). *A passion for excellence: The leadership difference.* New York: Random House.

Peters, T., & Waterman, R. (1982). *In search of excellence: Lessons from America's best-run companies.* New York: Harper & Row.

Rothstein, L.R. (1995). The empowerment effort that came undone. *Harvard Business Review, 73*(1), 20-30.

Simons, R.L., & Weston, H.A. (1991). *Nordstrom: Dissension in the ranks?* Harvard Business School Case 9-191-002.

Stewart, G.B. et al. (1993). Rethinking rewards. *Harvard Business Review, 71*(6), 37-49.

Sundstrom, E., DeMeuse, K.P., & Futrell, D. (1990). Work teams: Applications and effectiveness. *American Psychologist, 45*(2), 120-133.

Takahashi, D. (1996, September 12). Motorola sees weak profit in 3rd quarter. *Wall Street Journal.*

Therrien, L. (1989, November 13). The rival Japan respects—Motorola's secrets: Strong R&D, built-in quality, and zealous service. *Business Week,* 108-110, 114, 118.

Thompson, P.H., & Dalton, G.W. (1970). Performance appraisal: Managers beware. *Harvard Business Review, 48*(1), 149-157.

Welch, J. (1988, April). *Managing for the nineties.* General Electric Annual Meeting, Waukesha, WI.

Wetlaufer, S. (1988). *The anatomy of a "team destroyer": An analysis of individuals who stymie interfunctional coordination.* Harvard Business School Case Note 9-589-038.

Wetlaufer, S. (1994). The team that wasn't. *Harvard Business Review, 72*(6), 22–38.

AN INTEGRATIVE THEORETICAL FRAMEWORK FOR UNDERSTANDING TEAM REWARD ALLOCATION PREFERENCES

Anthony G. Parisi and Lillian T. Eby

ABSTRACT

This integrative theoretical framework for understanding team reward allocation preferences draws from research and theory in the areas of team development (Gersick, 1988; Tuckman, 1965; Tuckman & Jensen, 1977), time and work pacing (Kanfer & Ackerman, 1989; McGrath & Kelly, 1986), and equity versus equality norms for reward allocation (Leventhal, 1976). The framework proposes that an understanding of the role of time in team behavior may be especially important in regard to team performance management and reward allocation preferences. Team reward allocation preferences change over the lifespan of the team as a function of time-based variables. This includes changing task demands and shifting goals of the team over time. It is proposed that in the early stages of the team's lifespan less attention will be given to task-related concerns and members will prefer equality-based reward allocations. However, as the deadline for task completion

Advances in Interdisciplinary Studies of Work Teams, Volume 6, pages 161-186.

ISBN: 0-7623-0655-6

approaches, attention will shift to on-task activities, and there will be a corresponding change in preference toward equity-based reward allocations. Further, this change in reward allocation preference later in the lifespan of the team is expected to be moderated by task interdependence, cohesion, and team members' collectivistic orientation. Directions for future research and practical implications are discussed.

INTRODUCTION

Today's organizations have dramatically increased their use of work teams. Based on a recent survey of Fortune 1000 companies, 68 percent reported that they used self-managed work teams and over 90 percent used some form of employee participation teams (Lawler, Mohrman, & Ledford, 1995). Similarly, a 1992 survey found that 82 percent of U.S. companies reported that some of their employees worked in teams (Gordon, 1992). As companies use teams more extensively as primary work units, they are being implemented for many different purposes. This includes self-managed work teams, cross-functional project teams, executive teams, quality teams, advice and employee involvement teams, and customer service teams (Cohen & Bailey, 1997; Sundstrom, DeMeuse, & Futrell, 1990). Furthermore, while once used primarily in manufacturing settings, teams are becoming commonplace in a variety of industries and sectors (Eby, Adams, Russell, & Gaby, 1996). Given the central role of work teams in many organizations, research is needed to help companies design effective work teams and align human resource management (HRM) systems to support their use.

One of the most pressing HRM issues is how to design compensation systems in team-based organizations. Traditional individually based compensation practices such as merit pay, skill-based pay, and seniority systems may not be well-suited for team-based work, particularly if it is difficult to measure and assess individual contributions to team performance. Individually based reward systems may also undermine many firms' rationales for implementing teams, namely to increase collaboration and enhance efficiency (DeMatteo, Eby, & Sundstrom, 1998). This may occur because individually based rewards may create a competitive climate among team members, which in turn can hinder cooperation and team effectiveness (Gomez-Mejia & Balkin, 1992).

Due to these concerns about individually based reward systems, many organizations are experimenting with team-based rewards as a way to reinforce teamwork and facilitate team effectiveness. Descriptive reports of current compensation practices illustrate this growing trend. The American Productivity Center found that around 15 percent of the organizations surveyed used small team bonus systems for work units and autonomous work teams (O'Dell, 1989). Another survey-based effort by Lawler and colleagues, (1995) found that 70 percent of the Fortune 1000 companies surveyed reported the use of some form of team reward. Further, Lawler and colleagues (1995) found that 73 percent of companies surveyed reported the intention to increase their use of team rewards in the future.

While many organizations are using team-based reward systems, little research has examined the conditions under which team-based rewards are likely to be most effective in organizational settings (for further discussion see DeMatteo, Eby, & Sundstrom, 1998). In contrast, laboratory-based research using teams consisting of strangers working on team tasks for short durations of time has provided considerable insight into individuals' reward allocation preferences and the outcomes associated with different reward strategies (for reviews see Cotton & Cook, 1982; Johnson, Maruyama, Johnson, Nelson, & Skon, 1981; Miller & Hamblin, 1963). Broadly speaking, results of these studies show that when the team task is interdependent, rewards allocated using a cooperative strategy (e.g., equality based) lead to higher cooperation and productivity than rewards allocated using a competitive strategy (e.g., equity based). Recent research has continued to find similar results (for review see DeMatteo et al., 1998). Unfortunately, the artificial nature of laboratory settings and the short-term nature of these team tasks make generalizations to intact work teams in organizations somewhat tentative (DeMatteo et al., 1998). Furthermore, these studies typically do not take into consideration the possibility that reward allocation preferences change over the lifespan of the team, or that variables other than task interdependence may moderate reward allocation preferences. Due to these limitations, attention needs to focus on understanding team members' reward allocation preferences and the conditions under which team-based rewards are an effective team compensation strategy.

To offer insight into this important and timely issue, as well as provide practical suggestions to organizations implementing work teams, this chapter presents an integrative framework for understanding team members' reward allocation preferences. Rather than taking a static perspective of allocation preferences like most, if not all, of the previous research, we propose that reward allocation preferences are dynamic and changing over the lifespan of the team. This temporal perspective is consistent with much of the current research on team development, team effectiveness, and motivation theory (e.g., Gersick, 1988, 1989; Kanfer & Ackerman, 1989). By taking the concept of time into consideration in the study of team reward allocation preferences, this chapter also responds to researchers' call to incorporate temporal issues into the study of team-based phenomenon (Bettenhausen, 1991; Bluedornn & Denhardt, 1988; Moreland & Levine, 1988).

The proposed framework suggests that reward allocation preferences vary not only as a function of time, but also based on characteristics of the team and its members. While there are undoubtedly a myriad of factors that may moderate team reward allocation preferences, we focus on three variables which previous research has consistently identified as important in understanding a variety of team phenomenon: task interdependence, team cohesion, and team members' collectivistic orientation. Moreover, an attempt is made to isolate when in the lifespan of the team such factors are most likely to influence individuals' reward allocation preferences.

In addition to guiding future empirical research, studying team reward allocation preferences over time has important practical implications. For example, if team members' preferences change as a function of time, organizations may want to stagger the introduction of team reward systems to complement these changing preferences. In addition, if reward allocation preferences shift as a function of not only time, but also task interdependence, team cohesion, and the collectivistic orientation of team members, specific recommendations emerge to help organizations develop team compensation systems to match the unique characteristics of their work teams.

Before introducing the proposed framework for understanding team reward allocation preferences, it is important to define several key terms, outline underlying assumptions, and articulate boundary conditions associated with the framework. This will provide a common frame of reference for describing the proposed framework. Through this discussion, a brief synopsis of the proposed framework is provided. Next, the underlying theories germane to understanding the framework are reviewed in detail to illustrate the links between the proposed framework and existing theory and research. Having described the theoretical basis of the framework, specific components of the framework are explained. This is followed by a discussion of practical implications, limitations, and finally, an agenda for future research.

A FRAME OF REFERENCE FOR UNDERSTANDING TEAM REWARD ALLOCATION PREFERENCES

For purposes of the present chapter, the following definition of a work team is adopted: "an (interdependent) collection of individuals who share responsibility for specific outcomes for their organizations" (Sundstrom et al., 1990, p. 120). Generally speaking, this definition refers to teams that may be engaged in a variety of activities, including producing products, delivering services, making recommendations for improvements, developing new products or services, or strategic decision making (Cohen & Bailey, 1997). With this definition in mind, several boundary conditions are applicable to the proposed framework.

First, an underlying assumption of the framework is that there will be a starting point when the team is formed, and an end point when the team completes its tasks. This notion of starting and ending points is critical to understanding how team reward allocation preference may develop and change over time. As such, the proposed framework is useful only insofar as one can track and document such preferences over the lifespan of the team.

Second, it is also assumed that the members of the team will have competing demands for their time. It is expected that especially in the early stages of team formation, members will have responsibilities other than the team task. For example, a research team is likely to be working on multiple projects simultaneously, some of which may be individually based. Moreover, in teams such as data processing teams, there may be multiple demands for individuals' time (e.g., compiling individual work, sharing that work with team members, creating a final product). Thus, this framework is particularly applicable to research and development teams, cross-functional teams, long-standing quality teams, and long-standing project teams.

While it is expected that team members will have many different responsibilities, it is important that members perceive their role on the team as a central component of their job. Otherwise, the valence of the team reward system may not be strong enough to be of importance to team members. This leads us to the third boundary condition for the framework, that there must be some type of deadline for task completion. As Gersick (1988, 1989) notes, deadlines appear to be important cues for teams, and in the absence of such cues the team processes that develop may be quite different. As such, traditional production teams with little autonomy over deadlines and work schedules may not be applicable to the proposed framework.

Finally, the framework does not include teams that McGrath (1984) classifies as short-term task forces. These types of teams are characterized as being constrained in terms of their temporal scope and are typically in existence for a very short period of time. The current framework considers time to be an important component in understanding reward allocation preferences. Therefore, in the absence of time-based activities, the proposed framework is not expected to be applicable.

For both clarity and parsimony, the framework and its associated propositions are divided into two distinct time periods: Time 1 and Time 2. The underlying rationale for this demarcation of time is based on Gersick's (1988, 1989) studies of team development. Gersick provided evidence that the midpoint of a team's life is an important milestone after which the focus of the team shifts dramatically. We adopt this notion of a midpoint shift to illustrate how reward allocation preferences may change over the lifespan of the team. The proposed reason for this shift in reward allocation preference involves the goals which are most salient at these two points in time. Adopting Kanfer and Ackerman's (1989) notion of distal and proximal goals, it is proposed that during Time 1 completion of the team task is a distal goal whereas completing other (non-team) task duties, building solidarity, and developing interpersonal relationships with team members are proximal goals. These proximal goals at Time 1 should lead to equality-based reward allocation preferences.

In contrast, after the midpoint of the team (Time 2), distal and proximal goals are altered. In Time 2, the team task becomes a proximal goal due to the approach-

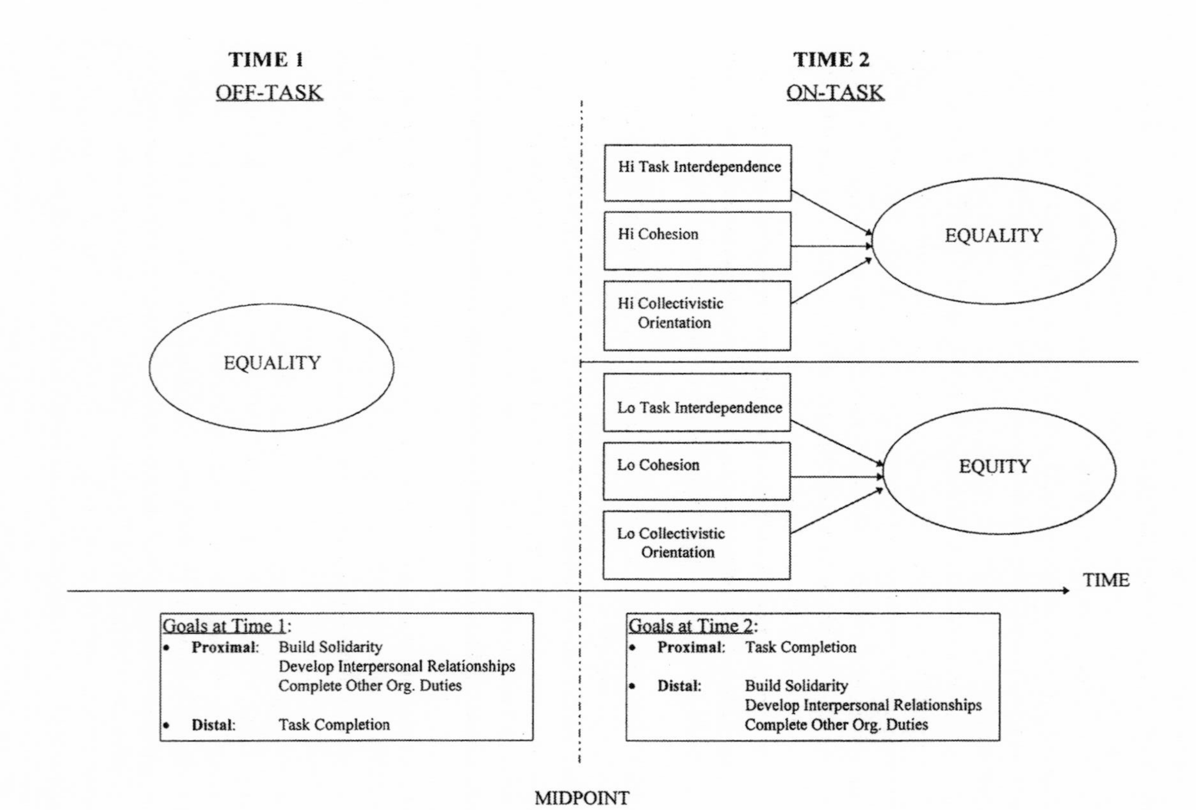

Figure 1. A Model of Team Reward Allocation Preferences

ing deadline, whereas other (non-team) tasks and building relationships with team members become distal goals. With these changes in goal salience, preferences for the type of team reward allocation strategy are expected to shift. Generally speaking, it is expected that members will express a preference for more equity-based reward allocation due to greater differentiation of member's contribution to task completion. However, the extent of this shift in allocation preference at Time 2 is expected to vary as a function of how interdependent the team's task is, how cohesive the team is, and the extent to which individuals on the team have a collectivistic orientation (see Figure 1).

Underlying Theoretical Orientations

Team Development

The research on team development provides a starting point for understanding how individuals' reward allocation preferences may develop over time. Unlike much of the empirical research on teams, a central premise of the team development literature is that teams are dynamic and changing entities, and that the nature of the team experience differs dramatically over the lifespan of the team (Gersick, 1988, 1989; Moreland & Levine, 1982, 1988; Tuckman, 1965; Tuckman & Jensen, 1977; Worchel, Coutant-Sassic, & Grossman, 1992). This suggests that a more complete understanding of teams will result from studying teams over an extended period of time.

Traditional models of group development have typically recognized that teams progress through phases as they work on a task. Tuckman and colleagues (1965, 1977) synthesized the group development literature and proposed a model which states that groups progress through a fixed hierarchy of stages which is the same for every group: forming, storming, norming, performing, and adjourning. In the forming stage, individuals come together around a task. The storming stage is characterized by conflict over how the task will be accomplished. During the norming stage, members agree upon ways to perform the task. Task completion characterizes the performing stage. Finally, the group disbands during the adjourning stage. Thus, traditional models of group development suggest that teams are initially concerned with establishing a clear identity and ensuring team cohesion and uniformity, yet in the later stages they divert their attention to productivity concerns and meeting the needs of individual members (Moreland & Levine, 1988; Tuckman, 1965; Tuckman & Jensen, 1977). In this traditional team development paradigm, the team's environment may constrain its ability to develop, but it will not alter the nature or sequence of the team's developmental stages.

Unfortunately, such traditional models provide snapshots of teams at different points in their lifespan, but say little about the mechanisms of change, how change is triggered, or how long a team will remain in any one stage (Gersick, 1988). Furthermore, many existing models have treated teams as closed systems (Goodstein & Dovico, 1979). That is, traditional models have failed to describe the role of the team's environment in its development. This lack of guidance regarding the interaction between environmental contingencies and teams' development result in the traditional models having limited utility for teams in organizations (Gersick, 1988). It usually is the case that tasks, resources, and requirements for success originate from outside the team (Gladstein, 1984). Therefore, the teams' ability to communicate with their environment will often be critical for its success (Gersick, 1988).

To overcome some of the limitations of traditional team development models, Gersick (1988) proposed that attention should focus on assessing the role of time, as well as the role of the environment, over the lifespan of the team. Gersick's (1988) principal research question was: "What does a group in an organization do, from the moment it convenes to the end of its lifespan, to create the specific product that exists at the conclusion of the last meeting?" (p. 11). Gersick examined teams from six different organizations and one university. Counter to traditional models, Gersick found that teams used widely diverse behaviors to accomplish their work, however the timing of when teams formed, maintained, and changed the way they worked was highly similar across teams.

Based on the results of a series of studies, Gersick (1988, 1989) described team development as a "punctuated equilibrium." During the team's first meeting, a framework of behavioral patterns and assumptions through which a team approached its project emerged, and the team maintained that pattern of behavior through the first half of its life. Further, Gersick found that teams often showed little visible progress during this time, because members may be unable to perceive a use for the information they were generating until they revise their initial behavioral routine. However, halfway through the life of the team, a paradigmatic shift in the team's approach to its work occurred, enabling team members to capitalize on the gradual skills and knowledge they accumulated to make significant advances toward goal achievement. This midpoint transition was a powerful opportunity for a team to alter its course, develop new strategies, and refocus its goals. However, Gersick also found that this transition must be used well, for once the midpoint passed teams were unlikely to alter their basic strategy again. The second half of the team's life was spent implementing the plans that crystallized during the transition. Gersick notes that from the midpoint on, team performance will be at its highest level, because attention has shifted toward task completion.

Gersick's (1988, 1989) findings illustrate that teams do not develop in a universal sequence of activities or stages, as traditional models predict. Rather, teams appear to establish perceptual and behavioral patterns very quickly and work within those patterns for relatively long periods. Teams appear to move toward task completion through concentrated bursts of revolutionary change in established behavioral patterns. The most striking single feature of Gersick's model, and possibly the most critical for understanding the development of reward allocation preferences in team-based settings, is the midpoint transition. In most cases, the midpoint was the most likely time at which team members felt both ready and willing to declare that it is time to move ahead with the work of the team. The special power of the midpoint appears to be that it is the most likely time that members will select, consciously or unconsciously, as a key milestone. Thus, teams are more likely to have an assembly of members simultaneously ready to make changes at that point than either earlier or later. Further, members are less likely to feel it is the wrong time to make changes than at any other point. Not only do teams tend to hurry more as the deadline approaches, but their struggles to fit work into the remaining time may become more particularistic as they deal with the task demands associated with the specific plans generated by the team (Gersick, 1989). Based on this pattern, it appears as if early on in the life of the team the distal deadline of task completion appears to be of little salience to the team. However, the passage of time may increase the salience of the team's deadline, which in turn, influences the manner on which team members allocate their time and energy (Seers & Woodruff, 1997).

In summary, some research suggests that the midpoint of a team's life is the signal for action. Prior to the midpoint, teams are less likely to concentrate their efforts to accomplish their respective tasks. However, after the midpoint, teams appear to allocate an increasing amount of time to task completion until they reach their team deadlines. Therefore, we propose that like other team phenomenon, the midpoint may serve as a critical time when members alter their behavioral routines and reassess their preferences for reward allocation within the team. Central to understanding why the midpoint may trigger changes in reward allocation preferences is a careful examination of what is likely to be occurring during the pre-midpoint (Time 1), compared to the post-midpoint (Time 2). The section that follows further describes the team behavior patterns that are expected during the two time periods.

Time and Work Pacing

Studies have shown that teams adjust both their rate of work (Gersick, 1988; McGrath & Kelly, 1986) and style of interaction (Frye & Stritch, 1964; Hermann, 1972; Pruitt & Drews, 1969) in response to time constraints. For example, McGrath and Kelly found that "the shorter the time limit on [an initial] work period, the higher the rate" at which teams studied in a laboratory solved anagrams (1986, p. 98). Teams given more time in their first trials not only worked more slowly but showed more attention to interpersonal matters in their interaction patterns (McGrath & Kelly, 1986). Other studies have shown that even under very structured working conditions, people find many opportunities to interject off-task activities into the work day (Abramis, 1989; Mars & Nicod, 1984; Roy, 1955, 1960; Terkel, 1972).

Recent developments in work motivation theory provide some insight into the reasons for these work-pacing effects. Several researchers suggest that behavior at work is a result of a resource allocation process individuals engage in involving the distribution of time and effort (Atkinson & Birch, 1974; Fichman, 1988; Kanfer, 1990; Naylor, Pritchard, & Ilgen, 1980). Individuals have limited amounts of attention and effort, and resource allocation theories propose that people distribute their limited time and effort across a wide variety of activities. Therefore, the distribution of effort (or "resources") depends on competing demands on individuals' time. As Kanfer and Ackerman (1989) suggest, individuals may allocate more time and resources to proximal goals rather than distal goals because resource demands for proximal goals are higher than corresponding demands for distal goals. Furthermore, such theories suggest that the allocation of time and resources to distal goals will steadily increase as distal goals become more proximal. For instance, an advertising work team may have many accounts, each with different deadlines. The team would most likely see the closest deadlines as proximal goals and other long-term deadlines as distal goals. However, as the deadline for an account that was at one time considered a distal goal nears, more attention and resources will be devoted to competing that task. Therefore, all goals, proximal and distal, vary in perceived importance based on time.

Based on these assumptions, if an individual chooses to engage in a particular behavior, then that individual also makes a corresponding choice not to engage in other behaviors (Ilgen, Dass, McKellin, & Lee, 1995). In the context of work, there are on-task (team-related) behaviors such as diagnosing a team problem, proposing solutions, and following through on the proposed course of action, and off-task (non-team-related) behaviors such as informal socializing on the job or engaging in other role-required activities not associated with the team task (Ilgen et al., 1995; Mullen & Cooper, 1994). The following example illustrates this resource allocation process whereby decisions or choices are made about how to spend one's limited time and cognitive resources. Consider a research team working on multiple projects simultaneously. The team will likely choose to perform task-related behaviors for one or maybe two projects at a given time, but not all projects simultaneously because of limited resources. Therefore, the research team may set priorities based on project deadlines and perform on-task behaviors related to those projects with the highest priorities (i.e., closest deadlines).

In summary, during the first half of a team's life, the team may not be focused on the task due to other activities outside of the team that require the time and effort of team members. Therefore, the first half of a team's life may be spent performing more off-task (non-team) behaviors than on-task (team) behaviors. Consistent with traditional models of team development (i.e., forming and storming stages), in the early stages of the team's development, team members are likely to be concerned with establishing a clear identity and ensuring team cohesion and uniformity (Tuckman, 1965; Tuckman & Jensen, 1977). These team building actions are not directly tied to task completion, rather they facilitate team member interaction. In contrast, the second half of the team's life is likely to be spent performing on-task behaviors (i.e., norming, performing) because the goal of task completion has become more proximal.

Initially, team members may perceive outcomes associated with off-task behaviors (e.g., fulfilling social needs and increasing job satisfaction) to be more attractive because those outcomes have a greater likelihood of occurring than outcomes related to the team task (e.g., praise for completing the task and avoidance of failure). Outcomes associated with the team task are perceived as having a weak link to valued rewards because on-task performance is a distal goal. However, as the deadline approaches, team members may realize that task completion will provide more attractive outcomes than off-task behaviors. For example, if the research team does not meet the deadline for completing its assigned project, team members will be reprimanded. In effect, if the task is not completed by the deadline, there may be negative outcomes which are more salient than the outcomes associated with off-task behaviors. The next section will elaborate on the relationship among off-task and on-task behaviors with reward allocation preferences.

Equity versus Equality Norms for Reward Allocation

Research has shown that there are conditions under which individuals proactively employ various justice norms (Greenberg & Leventhal, 1976). Justice norms refer to appropriate strategies for distributing goods and outcomes among individuals within a team and are believed to impact the well-being of individual team members as well as the team as a whole (Deutsch, 1975). Two distinct allocation rules that may be used to distribute rewards within a team are the equality norm and the equity norm. The equality norm states that rewards will be divided equally regardless of differential contributions among members. In contrast, the equity norm states that rewards will be distributed proportional to the contributions of individual team members.

According to Leventhal's (1976) justice judgment model, individuals attempt to make fair allocation decisions by applying these allocation rules to the situations they confront. The choice of a preferred allocation strategy is often dictated by situational factors (Deutsch, 1975; Leventhal, 1976) with different norms of justice followed under different circumstances. For instance, when the importance of maintaining solidarity among team members is critical, the allocation procedure viewed as "fair" by team members may be the use of an equality norm (Deutsch, 1975). However, if the major concern is performance or productivity, then reward allocations based on the equity norm would be preferable (Deutsch, 1975, 1985; Miller & Komorita, 1995).

This literature suggests that in general, equal distribution of rewards is more likely to foster team harmony and cooperation among team members. Thus, members of new teams may be satisfied with an incentive system that provides an equal distribution of rewards, at least until the team has worked out its members' roles and working relationships (i.e., progressed through the storming stage). Furthermore, team members may prefer the equality allocation norms because most of the behaviors associated with the first half of the team's life are likely to be off-task behaviors (Gersick, 1988, 1989; Tuckman, 1965; Tuckman & Jensen, 1977). Team members may not feel that it is fair to be rewarded differentially during these early stages of the team development because they may not have performed many behaviors related to the completion of the team task. However, as the team develops over time, members may respond differently to team-based rewards allocated using the equality norm. As the team matures, members may notice differential contributions by individuals in the team. As such, allocation preferences may shift to that of an equity-based norm to reflect those contributions. In effect, team development factors may give rise to a norm shift in the team from one of equality to one of equity. However, it should be noted that such a norm shift may be more likely when task interdependence is low, cohesion is low, and members have more of an individualistic orientation. Gersick's research, coupled with the research on time and work pacing, suggest that this norm shift is most likely to occur at the midpoint of the team's life and perhaps grow stronger as the goal of task completion becomes more proximal.

Taken together, this research suggests that individuals have a limited amount of resources to allocate to various team-related and non-team related activities. Further, off-task and on-task behaviors may be related to team development and reward allocation preferences. During the first half of the team's life, it is expected that team members will perform predominantly off-task behaviors and will prefer to be rewarded based on the equality norm. At the midpoint, the team will realize that it is time to concentrate on the task in order to complete it prior to the deadline. Therefore, the second half of the team's life will consist of team members performing on-task behaviors. The execution of these on-task behaviors are expected to cause a shift to set the stage for a potential norm shift from an equality to equity reward allocation preference.

An Integrated Approach

While time appears to play an important role in understanding both individual motivation and team behavior, existing research on teams has been criticized for its general failure to take into account the importance of time and context in understanding team phenomena (Argote & McGrath, 1993; Goodstein & Dovico, 1979). However, an understanding of the role of time in team behavior may be especially important in regard to team performance management and reward allocation preferences. As illustrated in Figure 1, over the course of the lifespan of the team, different goals are likely to be salient and proximal to team members. Additionally, regardless of how work is arranged, team members are likely to have competing pressures for their time and energy. Therefore, team members' choice to exert effort on a particular task depends on the salience of the task to the team, as well as the proximity of the goal associated with that particular task. This suggests that different reward practices may be necessary to foster motivation, and ultimately performance, over the lifespan of the team.

Having described the theoretical basis of the framework, the section that follows will explain and elaborate on the specific components of the framework presented in Figure 1. The team is formed to complete a certain task at Time 1. At this point, proximal goals are expected to be building harmony and developing interpersonal relationships among team members. Furthermore, the completion of other organizational duties (off-task team behaviors) are expected to represent proximal goals. In organizational settings, team members typically perform a variety of tasks and have various affiliations that force them to pace their use of a limited resource, time, in order to meet multiple task deadlines. Thus, regardless of how work is arranged, individuals appear to have competing pressures for their time and energy. Team members must continually make decisions about how to allocate their time and energy among different tasks. Therefore, proximal goals during time 1 lead to team behavior which is not oriented toward the team's task (off-task). As such, task completion is considered a distal goal during Time 1 (see Figure 1).

It is expected that the reward allocation procedure which will facilitate the achievement of these proximal goals at Time 1, as well as be viewed as fair by team members, is the equality norm. Under the equality norm, rewards will be divided equally among team members regardless of the individual contribution of specific members. The rationale for this expectation is that the equality norm has been associated with maintaining team harmony and minimizing team conflict (Miller & Komorita, 1995). Thus, in the early stages of team formation when interpersonal relationships are developing, roles are unclear, individual contributions to the team task are less distinct, and there is less on-task behavior, an equality-based reward allocation may be preferred (see Figure 1). Such a reward allocation practice is likely to foster perceptions of oneness and cohesion, and reduce initial apprehensions commonly associated with working in teams such as concerns about free-riding and social loafing (Shepperd, 1993; Tjosvold, 1984). Laboratory-based research on the effect of competitive versus cooperative reward structures supports the contention that cooperative rewards foster more positive interpersonal relationships and cooperation among team members working on a team-oriented task (for review see DeMatteo et al., 1998).

In contrast, as the team begins to focus its attention on the demands of the task, the amount of time spent on off-task behavior is expected to decrease and on-task behavior is expected to increase (Gersick, 1989) (see Figure 1). At Time 2 the team has passed the midpoint of its existence. Thus, consistent with Gersick's model, the team should refocus its goals toward task completion. That is, the proximal goal becomes task completion and the distal goals become building solidarity, developing interpersonal relationships among team members, and completing other individual responsibilities. It is expected that during Time 2, the team will steadily increase its on-task behaviors and reduce its off-task behaviors until task completion. As the on-task behaviors increase, team members may be able to better differentiate individual contributions and become more concerned with the fairness of team reward allocations. Team members may feel that each member deserves their unique contribution to team performance to be tied directly to the amount of compensation they receive. This is consistent with expectancy theory predictions which state that as the association between individual effort and individual performance, or the link between individual performance and individual outcomes grows stronger, motivation and performance will increase (Porter & Lawler, 1968; Vroom, 1964). Thus, maintaining a clear contingency between individual and team members' contributions and rewards may heighten motivation and performance, increase perceptions of fairness, and decrease the likelihood of unequal contributions to team performance (Deutsch, 1975, 1985; Karau & Williams, 1993; Miller & Komorita, 1995; Shepperd, 1993). In summary, this suggests that it may be important to adopt equity-based reward allocation procedures at Time 2 to ensure that individual contributions are rewarded and that accountability within the team is maintained.

Taken together, it appears as if the reward allocation norm of equity may be perceived as more suitable once the team begins focusing their effort on task completion (Time 2). An allocator may adopt the equity principle of allocation rather than the equality principle because adopting the equity principle has certain desired effects (Leventhal, 1976; Miller & Komorita, 1995). For instance, numerous studies have shown that equitable reward allocation can facilitate task performance. That is, providing large rewards to good performers may be a motivator to individuals seeking large rewards and, therefore, equitable allocation most likely will maximize performance (Leventhal, 1976; Miller & Komorita, 1995). In contrast, perceptions of unfairness based on the type of allocation principle may negatively impact affective reactions. For example, perceptions of unfairness regarding the allocation principle may result in decreased satisfaction with pay as well as increased intentions to leave the organization (e.g., Adams, 1965; Moorman, 1991). This suggests that it may be to the organization's advantage to ensure that the proper reward allocation principle is utilized. While there is theoretical and empirical support for the expected shift in reward allocation preferences as the deadline for task completion approaches, several variables are proposed as moderators of the relationship between time and reward allocation preferences.

Proposed Moderating Variables

There are several post-midpoint variables which may moderate the relationships illustrated in Figure 1. These moderating variables include task interdependence, team cohesion, and team members' collectivistic orientation. It should be noted that this list of moderating variables is not an exhaustive list. It is possible that other moderating variables exist, however the aforementioned variables will be the focus of the present discussion. The selection of these variables was driven by previous research in the area of work team behavior.

Task Interdependence. Perhaps the most important variable in the current framework is the nature of the team task. Saaverdra, Earley, and Van Dyne (1993) reviewed the literature on task interdependence and concluded that tasks vary regarding the interdependence required by team members. Research suggests a hierarchy of task interdependence reflecting increasing levels of dependence among individuals based on the exchange of information or resources: pooled, sequential, reciprocal, and team methods of exchange (Thompson, 1967; Van de Ven, Delbecq, & Koenig, 1976). Pooled interdependence or independent work flow is defined as a situation in which each team member contributes to the team output without the need for direct interaction among team members. Team performance is reflected in the sum of individual team members' performances. Sequential interdependence or one-way work flow occurs when one team member must act before another team member can act. Team performance is reflected by each team member performing successfully and in the correct order. Reciprocal interdependence or two-way work-flow occurs when one team member's output is another member's input and vice versa—temporally lagged, two-way interactions (Van de Ven et al., 1976). In this type of task interdependence, team performance is reflected in the team's ability to coordinate among members. Finally, team interdependence is characterized by a simultaneous, multi-directional work flow. That is, team members work together to diagnose and solve problems and collaborate to complete a task (Saaverdra et al., 1993).

Research indicates that task interdependence impacts a variety of team attitudes and behaviors (Saaverdra et al., 1993; Shea & Guzzo, 1987; Wageman & Baker, 1997), and may impact the effectiveness of team-based rewards (DeMatteo et al., 1998). Thus, it is expected that when task interdependence is very high among team members (i.e., team interdependence), the preference for equity-based reward allocation will be less strong at Time 2 due to the coordinated effort required for task completion and the difficulty differentiating individual contributions. Teams functioning under team interdependence may involve a large number of interactions between team members in order to exchange information and resources necessary for task completion (Saaverdra et al., 1993). This almost constant interaction among team members blurs team member roles and produces difficulty in recognizing individual differences in performance. Therefore, under conditions of high task interdependence (i.e., reciprocal or team interdependence), an equality-based reward allocation may be preferable in order to reward the team for performing as a team.

Cohesion. Team cohesion is also proposed as moderating the strength of team members' preferences for equity-based reward allocation at Time 2. Cohesiveness has been defined as the attraction of members to their team, and develops based on liking among team members, attraction to the team task, and attraction to the status associated with the team (Cartwright, 1968; McGrath, 1984). In the context of reward allocation preferences, it is expected that when team cohesion is high at Time 2, members' preferences for equity-based reward allocations will be less pronounced. The logic is that highly cohesive teams may be more concerned about the effect that moving to a more individually oriented reward system (i.e., equity-based reward allocation) may have on existing levels of team cohesion than less cohesive teams. This is expected because highly cohesive teams may prefer to maintain harmony and minimize conflicts among team members (McGrath, 1984). The team's sense of cohesion should not be undermined by a shift in reward allocation. Therefore, although the goals shift toward task completion after the midpoint, a shift to equity-based allocation preference may be perceived as unfair to a highly cohesive team. Under these team conditions, equality-based reward allocation may be preferable to equity-based reward allocation.

Team Members' Collectivistic Orientation. An individual difference variable that may be central to understanding how an employee responds to working on a team is the team member's collectivistic orientation (Eby & Dobbins, 1997). Wagner (1995) defined collectivism as the condition in which the demands and interests of the team are accorded greater importance than the desires and needs of the individual. Whereas collectivists frequently prefer to cooperate in teams, individualists exhibit a tendency to avoid cooperation. Individualism occurs when personal interests are more salient than the needs of the team. Applied to reward allocation preferences among members of work teams, cooperative appeals based on the contingent receipt of collective outcomes (equality norm) are likely to have little or no effect on the behavior of individualists because collective outcomes must be shared with the other team members rather than personally consumed (Wagner & Moch, 1986). Therefore, preferences for equality-based reward allocations may be influenced by members' collectivistic orientation.

In fact, there is some empirical evidence that collectivists prefer reward allocation procedures that are based on equality rather than equity distribution rules (Leung & Bond, 1984). In a study of intact teams, DeMatteo and Eby (1997) found that collectivists were generally more satisfied with team-based rewards than individualists. In addition, Eby and Dobbins (1997) found that teams comprised of members who were more collectivistic were also more likely to engage in highly cooperative team behaviors. Further, Eby and colleagues (1996) determined that individuals reporting higher collectivistic tendencies were more receptive to a large-scale organizational change initiative involving the implementation of work teams. Taken together, it seems reasonable to propose that collectivists may have a tendency to engage in more cooperative team behavior, be more comfortable under team conditions where individual contributions are less salient, and have a general preference for more equality-based reward allocation procedures.

Given the proposed framework there are a number of practical implications as well as research gaps that need to be addressed. The following sections suggest future directions regarding the framework's utility.

PRACTICAL IMPLICATIONS

One of the most obvious practical implications has to do with the notion that team reward allocation preferences may change as a function of time. If preferences do shift over time, organizations may want to stagger the introduction of team reward systems to complement team members' changing preferences. Furthermore, the framework suggests that reward preferences may shift not only as a function of time, but also task interdependence, team cohesion, and the collectivistic orientation of team members. Thus, it may be advantageous for the organization to match compensation systems with the unique characteristics of the work team. However, recommendations stemming from the proposed framework may result in increased complexity in administering team reward systems. For instance, an organization may have a number of teams performing a variety of tasks at any one time. The amount of information and manpower necessary to ensure that teams are being monitored regarding deadlines for task completion, task interdependence, team cohesion, and team member collectivistic orientation may be overwhelming. Furthermore, the administrative duties necessary to control the rewards for all individuals that are members of a number of teams that are qualitatively different may be impossible. Further research may begin to provide guidance in applying components of the proposed framework in organizational settings.

While organizations may want to stagger the introduction of team-based rewards, it should be noted that team rewards are not meant to be the sole basis of compensation for individual team members. A mixed system of rewards in which a portion of the rewards are based on team performance and a portion related to individual performance is recommended. In practice, most organizations adopt a policy regarding the distribution of team-based rewards. O'Dell (1987) reports that while the dollar amount allocated in team-based reward systems is typically greater than that in organization-wide pay systems such as gainsharing, the average size of team-based rewards are between 10.3-12.2 percent of base pay. Further, a recent survey by DeMatteo, Rush, Sundstrom, and Eby (1997) asked employees to report the size of their most recent team-based bonus. While the mean was $928.43 or 3.06 percent of base pay, some teams received a bonus as large as $10,000, suggesting that the size of the team bonus varies greatly across organizations. It should also be noted that some teams probably did not reach their goals that year and did not receive a bonus.

Furthermore, within such a team-based system, allocation rules may differ such that an individual may receive a bonus based on the team's overall performance (i.e., equality-based allocation) or individual contribution to team performance (equity-based allocation). Under these conditions, a critical issue is the relative size of the team-based reward in comparison to the individual-based compensation. Specifically, if the size of the team-based reward is quite small in proportion to base pay (e.g., 5% of the individual's base salary), then the team bonus may not be significant enough to affect the individual's motivation or satisfaction (Patten, 1977). As such, it is recommended that if team-based rewards are to be utilized in organizational settings, the size of team rewards needs to be large enough to motivate and sustain team performance. Furthermore, consistent with research on employee participation programs (Goodman, Devadas, & Hughson, 1988), organizations may want to consider involving team members in the process of deciding if, and when, the team should switch from equality to equity-based reward allocation in order to increase work attitudes, motivation, and performance.

Although this chapter has focused on shifting reward allocation preferences within the team, it should not be assumed that reward allocation preferences always change over time. As is noted in Figure 1, if task interdependence, team cohesion, and/or team members' collectivistic orientation is high at Time 2, then equality may be appropriate throughout the team's lifespan. Furthermore, if teams are responsible for tasks that are low on interdependence (i.e., pooled interdependence or sequential interdependence), perhaps members should be rewarded using an equity-based norm throughout the team's entire lifespan. Tasks low on interdependence may not require team cohesion and should be well-suited for allocation strategies that facilitate clear recognition of individual members' performance. As the law of effect states, behavior that is perceived to lead to rewards will tend to be repeated (Lawler, 1966), therefore, clear links should be made between performance and rewards. Equity-based reward allocation provides an unambiguous linkage between performance and rewards on tasks low on interdependence. However, the simplicity of the law of effect should not detract from the complexity involved in putting the law of effect into practice. Organizations should facilitate the perception that good performance leads to increased rewards. Nonetheless, forming and maintaining a link between performance and rewards is becoming more and more complicated as organizations move toward team-based work.

LIMITATIONS

While the proposed framework is one of the first to examine the complex relationships among time-based variables, individual and team characteristics, and team reward allocation preferences, it is not without its limitations. First, the proposed framework does not include organizational context factors that may influence individuals' receptivity to team-based reward systems in general, or specific preferences for the equality- or equity-based reward allocation norms. For instance, if an organization has a highly individualistic organizational culture, the use of team-based rewards in general, and particularly equality-based allocation procedures, may be viewed negatively by organization members. In turn, positive effects expected from equality-based rewards such as increased performance and satisfaction may not be forthcoming. Similarly, a collectivistic culture may influence the perceived appropriateness of equity-based reward allocation. In contrast, the use of equality-based rewards in a highly collectivistic culture may be viewed positively by organizational members.

Another organizational consideration may be the congruence of an organization's compensation system with the other human resource management systems within the organization. Recent research on strategic human resource management suggests that organizations may be most effective when individuals are selected for teams, trained on skills necessary for functioning on a team, evaluated on team performance, and compensated based on team performance (Arthur, 1994; Huselid, 1995; Youndt, Snell, Dean, & Lepak, 1996). The goal of strategic human resource management is to provide a clear and consistent message as to what is valued by the organization. Saaverdra and colleagues (1993) empirically examined the interrelationships between task, goal, and feedback interdependence, finding that congruency among all three led to higher levels of group performance. Failure to provide an unambiguous and consistent message may result in decreased organizational effectiveness (Youndt et al., 1996). Therefore, organizations may be limited in the types of team reward allocation procedures that would be appropriate due to existing human resource management systems. For example, if performance appraisal systems recognize and reinforce individual behavior rather than effective teamwork behavior, team-based reward systems may be ineffective, because the behavior being evaluated is different from that which is being rewarded.

In addition, all team members may not be satisfied with the reward allocation norm that is adopted within the team. Task interdependence and team cohesion are presented as stable characteristics of the team and, thus, are not expected to lead to differences in individual team members' reactions to the reward allocation norm that is adopted. However, teams are likely to be composed of members with varying collectivistic orientations, making it difficult to meet all members' reward allocation preferences. This suggests that organizations may need to form homogeneous teams based on members' collectivistic orientations (high or low) in order to increase the likelihood that all team members will respond favorably to the reward allocation norm that is adopted.

Furthermore, previous research has illustrated that individuals' collectivistic orientations may be altered over time (Breer & Locke, 1965). For instance, if individualistic team members are initially rewarded based on the equality norm, they may learn to value equality which may, in turn, influence subsequent reward allocation preferences. In sum, although we expect that members' collectivistic orientations may play a critical role in reward allocation preferences, the specific nature of its effect remains somewhat unclear. Future research is needed to examine how teams with varying levels of collectivism among its members regard team-based reward preferences.

A final limitation is that although the framework presented was based on previous research in a number of theoretical areas, most of the evidence is based on lab research. It has been suggested that lab research serves to enhance theory and provide optimal tests of research hypotheses (Driskell & Salas, 1992; Mook, 1983). However, the generality of lab results has been questioned (Brunswick, 1955; Campbell & Stanley, 1963), and replication in the field is a necessary requirement for the thorough testing of a theory (Driskell & Salas, 1992). As such, some of the links in the proposed theory that are based on laboratory research may be somewhat tentative. Furthermore, conclusions based on the relationships noted in Figure 1 await empirical confirmation. Specifically, longitudinal field-based research conducted is needed to examine the relationships depicted in Figure 1. Moreover, practical applications of the framework should proceed with caution until further research examining the framework is conducted. Although there are some limitations associated with the proposed framework, these limitations appear to be offset by the theoretical contribution to the literature on reward practices on work teams and the potential contribution to practice.

AN AGENDA FOR RESEARCH

The integrative framework in Figure 1 provides an initial step toward examining how team members' reward allocation preferences may develop and change over time. Many researchers have highlighted the lack of attention to the dynamic nature of teams and some have speculated that the omission of time as a variable in team-based research has hindered researchers' ability to fully understand the complexity of team-related phenomena (e.g., Bettenhausen, 1991). Due to its recognized importance, it is surprising that such little attention has been devoted to specifying how time-based variables may impact team members' reactions to performance management and reward systems. This chapter provides an initial examination of how the passage of time may impact members' reward allocation preferences due to the shifting nature of goals over the lifespan of the team.

By recognizing the importance of time in understanding reward allocation references, this chapter raises many research questions and provides several avenues for empirical research. This includes overarching research questions to more specific questions based on the framework proposed in the current chapter. Table 1 provides a summary of these research questions, organized into three main sections. The first section poses general research questions related to the applicability of the underlying theoretical orientations to the study of work teams. The second section addresses specific components of Figure 1. Finally, the third section goes beyond Figure 1 by proposing additional areas for future research to address some

Table 1. Questions to Guide Future Research

Section I: General Research Questions (RQ)	
RQ 1:	Does Gersick's (1988) finding of a punctuated equilibrium for teams performing a single task generalize to situations where a team members are responsible for multiple tasks?
RQ 2:	What are the specific behaviors that team members engage in during the off-task (pre-midpoint) time period? Are these behaviors relevant to building team solidarity, unrelated to the team's development (e.g., working on tasks associated with other roles and responsibilities), or a combination of both?
RQ 3:	When team members choose to engage in off-task (non-team-related) behaviors does the rate of on-task (team-related) behavior actually decrease or does the rate of off-task behaviors simply increase?
RQ 4:	When team members choose to engage in on-task (team-related) behaviors does the rate of off-task (non-team-related) behavior actually decrease or does the rate of on-task behaviors simply increase?
Section II: Specific Components of Figure 1	
RQ 5:	What do team members report as proximal goals prior to the midpoint of the team's existence?
RQ 6:	What do team members report as distal goals prior to the midpoint of the team's existence?
RQ 7:	What do team members report as proximal goals after the midpoint of the team's existence?
RQ 8:	What do team members report as distal goals after the midpoint of the team's existence?
RQ 9:	When does a proximal goal become a distal goal and vice versa?
RQ 10:	Do team members prefer team reward allocation practices that reflect equality at Time 1?
RQ 11:	Do team members prefer team reward allocation practices that reflect equity at Time 2?
RQ 12:	Are team members' preferences for team reward allocations at Time 2 moderated by task interdependence? If so, is the relationship such that under conditions of high task interdependence the preference will be an equality allocation strategy whereas under conditions of low task interdependence the preference will be an equity allocation strategy?
RQ 13:	Are team members' preferences for team reward allocations at Time 2 moderated by cohesion? If so, is the relationship such that under conditions of high cohesion the preference will be an equality allocation strategy whereas under conditions of low cohesion the preference will be an equity allocation strategy?
RQ 14:	Are team members' preferences for team reward allocations at Time 2 moderated by team members' collectivistic orientation? If so, is the relationship such that under conditions where members have a strong collectivistic orientation there will be a preference for an equality allocation strategy, whereas under conditions where members have weak collectivistic orientations the preference will be an equity allocation strategy?
RQ 15:	Is congruence among team allocation preference and reward allocation strategy related to important outcomes at the individual level (e.g., satisfaction, motivation, turnover intentions)?
RQ 16:	Is congruence among team allocation preference and reward allocation strategy related to important outcomes at the team level (e.g., team turnover rates, morale)?
Section III: Additional Areas for Future Research	
RQ 17:	If task interdependence is average at Time 2, what will the preferred team reward allocation strategy be?
RQ 18:	If cohesion is average at Time 2, what will the preferred team reward allocation strategy be?

(continued)

Table 1. (Continued)

RQ 19:	If members' collectivistic orientation is average at Time 2, what will the preferred team reward allocation strategy be?
RQ 20:	If some team members are high on collectivism whereas others are low on collectivism, what will be team members' reward allocation preference?
RQ 21:	Are team members' preferences for team reward allocations at Time 2 moderated by goal interdependence?
RQ 22:	Are team members' preferences for team reward allocations at Time 2 moderated by feedback interdependence?
RQ 23:	Given that individuals from different cultures differ in their level of individualism-collectivism (e.g., United States vs. Japan), are there culturally based boundary conditions on the proposed framework?
RQ 24:	Do the relationships depicted in Figure 1 change if the team is responsible for multiple, sequential task projects?
RQ 25:	What other individual, team, and organizational variables influence team members' reward allocation preferences?

of the limitations discussed previously. Using Table 1 as a guide, it is our hope that future research on team-based reward practices will maintain a careful balance between being firmly grounded in theory and providing practical guidance to those charged with the difficult issue of how to design team-based rewards systems that meet the dual, and often conflicting goals of motivating individual team members and encouraging team cooperation.

CONCLUSION

As organizations increasingly reorganize work around teams, decisions of how to manage team performance and compensate team members will be a growing concern. The framework proposed in this chapter provides a starting point for efforts to understand team members' reward allocation preferences and how these changes may evolve over the lifespan of the team. Using existing theory and research, the integrative framework illustrated in Figure 1 provides a foundation for empirical research aimed at uncovering the complex relationship between time-based variables, characteristics of the team and its members, and preferences for different team-based reward strategies. Future research exploring this important topic may emerge that helps organizations develop effective team-based compensation packages that reinforce and reward team members' contributions to team success.

ACKNOWLEDGMENT

The authors wish to gratefully acknowledge Jacquelyn DeMatteo for her helpful comments and advice.

REFERENCES

Abramis, D.J. (1989). Finding the fun at work. *Psychology Today,* 36-38.

Adams, J.S. (1965). Inequity in social exchange. In L. Berkowitz (Ed.), *Advances in experimental social psychology* (vol. 2, pp. 267-299). New York: Academic Press.

Argote, L., & McGrath, J.E. (1993). Group processes in organizations: Continuity and change. In C.L. Cooper and I.T. Robertson (Eds.), *International review of industrial and organizational psychology, 8,* 332-389.

Arthur, J.B. (1994). Effects of human resource systems on manufacturing performance and turnover. *Academy of Management Journal, 37,* 670-687.

Atkinson, J.W., & Birch, D. (1974). The dynamics of achievement-oriented activity. In J.W. Atkinson, & J.O. Raynor (Eds.), *Motivation and achievement.* Washington, DC: Winston & Sons.

Bettenhausen, K.L. (1991). Five years of groups research: What have we learned and what needs to be addressed. *Journal of Management, 17,* 345-381.

Bluedornn A.C., & Denhardt, R.B. (1988). Time and organizations. *Journal of Management, 14,* 299-320.

Breer, P.E., & Locke, E.A. (1965). *Task experience as a source of attitudes.* Homewood, IL: The Dorsey Press.

Brunswick, E. (1955). Representative design and probabilistic theory in a functional psychology. *Psychological Review, 62,* 193-217.

Campbell, D.T., & Stanley, J.C. (1963). *Experimental and quasi-experimental designs for research.* Boston: Houghton-Mifflin.

Cartwright, D. (1968). The nature of group cohesiveness. In D. Cartwright & A. Zander (Eds.), *Group dynamics: Research and theory* (3rd ed.). New York: Harper & Row.

Cohen, S.G., & Bailey, D.E. (1997). What makes teams work: Group effectiveness research from shop floor to the executive suite. *Journal of Management, 23,* 239-290.

Cotton, J.L., & Cook, M.S. (1982). Meta-analyses and the effects of various reward systems: Some different conclusions from Johnson et al. *Psychological Bulletin, 92,* 176-183.

DeMatteo, J.S., & Eby, L.T. (1997). Who likes team rewards? An examination of individual difference variables related to satisfaction with team-based rewards. *Academy of Management Proceedings* (pp. 160-163). Statesboro, GA: Georgia Southern University.

DeMatteo, J.S., Eby, L.T., & Sundstrom, E. (1998). Team-based rewards: Current empirical evidence and directions for future research. *Research in Organizational Behavior, 20,* 141-183.

DeMatteo, J.S., Rush, M.C., Sundstrom, E., & Eby, L.T. (1997). Factors related to the successful implementation of team-based rewards. *ACA Journal,* 16-28.

Deutsch, M. (1975). Equity, equality, and need: What determines which value will be used as the basis of distributive justice? *Journal of Social Issues, 31*(3), 137-149.

Deutsch, M. (1985). *Distributive justice.* New Haven, CT: Yale University Press.

Driskell, J.E., & Salas, E. (1992). Can you study real teams in contrived settings? The value of small group research to understanding teams. In R.W. Sweeny and E. Salas (Eds.), *Teams: Their training and performance* (pp. 101-123). Norwood, NJ: Ablex Publishing.

Eby, L.T., Adams, D.M., Russell, J.E.A., & Gaby, S.G. (1996, April). *Factors related to readiness for change to team-based sales.* Paper presented at the annual meeting for the Society for Industrial and Organizational Psychology, San Diego, California.

Eby, L.T., & Dobbins, G.H. (1997). Collectivistic orientation in teams: An individual and group-level analysis. *Journal of Organizational Behavior, 18,* 275-295.

Fichman, M. (1988). Motivational consequences of absence and attendance: Proportional hazard estimation of a dynamic motivation model. *Journal of Applied Psychology, 73,* 119-134.

Frye, R., & Stritch, T. (1964). Effect of timed vs. non-timed discussion upon measures of influence and change in small groups. *Journal of Social Psychology, 63,* 139-143.

Gersick, C.J.G. (1988). Time and transition in work teams: Toward a new model of group development. *Academy of Management Journal, 31*(1), 9-41.

Gersick, C.J.G. (1989). Marking time: Predictable transitions in task groups. *Academy of Management Journal, 32*(2), 274-309.

Gladstein, D. (1984). Groups in context: A model of task group effectiveness. *Administrative Science Quarterly, 29,* 499-517.

Gomez-Mejia, L.R., & Balkin, D.B. (1992). *Compensation, organizational strategy, and firm performance.* Cincinnati, OH: South-Western Publishing Co.

Goodman, P.S., Devadas, R., & Houghson, T.L.G. (1988). Groups and productivity: Analyzing the effectiveness of self-managing teams. In J.P. Campbell and R.J. Campbell (Eds.), *Productivity in organizations* (pp.295-327). San Francisco: Jossey-Bass.

Goodstein, L.D., & Dovico, M. (1979). The decline and fall of the small group. *Journal of Applied Behavioral Science, 15,* 320-328.

Gordon, J. (1992, October). Work teams: How far have they come? *Training,* 59-65.

Greenberg, J., & Leventhal, G.S. (1976). Equity and the use of over-reward to motivate performance. *Journal of Personality and Social Psychology, 34,* 179-190.

Hermann, C. (1972). Threat, time, and surprise: A simulation of international crisis. In C. Hermann (Ed.), *International crises: Insights from behavioral research* (pp. 187-211). New York: Free Press.

Huselid, M.A. (1995). The impact of human resource management practices on turnover, productivity, and corporate financial performance. *Academy of Management Journal, 38,* 635-672.

Ilgen, D.R., Dass, P., McKellin, R.A., & Lee, S.L. (1995). Off-task behaviours and outcomes: An explanatory investigation of a neglected domain in organisational behaviour. *Applied Psychology: An International Review, 44*(1), 57-76.

Johnson, D.W., Maruyama, G., Johnson, R., Nelson, D., & Skon, L. (1981). Effects of cooperative, competitive, and individualistic goal structures on achievement: A meta-analysis. *Psychological Bulletin, 89,* 47-62.

Kanfer, R. (1990). Motivation theory and industrial and organizational psychology. In M.D. Dunnette & L.M. Hough (Eds.), *Handbook of industrial and organizational psychology* (vol. 1, pp. 75-170). Palo Alto, CA: Consulting Psychologists Press.

Kanfer, R., & Ackerman, P.L. (1989). Motivation and cognitive abilities: An integrative/aptitude-treatment interaction approach to skill acquisition. *Journal of Applied Psychology, 74*(4), 657-690.

Karau, S.J., & Williams, K.D. (1993). Social loafing: A meta-analytic review and theoretical integration. *Journal of Applied Psychology, 65*(4), 681-706.

Lawler, E.E., III (1966). Managers' attitudes toward how their pay is and should be determined. *Journal of Applied Psychology, 50,* 273-279.

Lawler, E.E., III, Mohrman, S.A., & Ledford, G.E. (1995). *Creating high performance organizations: Practices and results of employee involvement and total quality management in Fortune 1000 companies.* San Francisco: Jossey-Bass.

Leung, K., & Bond, M.H. (1984). The impact of cultural collectivism on reward allocation. *Journal of Personality and Social Psychology, 47,* 793-804.

Leventhal, G.S. (1976). Fairness in social relationships. In J.W. Thibaut, J.T. Spence, & R.C. Carson (Eds.), *Contemporary topics in social psychology* (pp. 211-239). Morristown, NJ: General Learning Press.

Mars, G. & Nicod, M. (1984). *The world of waiters.* London: George Allen & Unwin Ltd.

McGrath, J.E. (1984). *Groups: Interaction and performance.* Englewood Cliffs, NJ: Prentice-Hall.

McGrath, J.E., & Kelly, J.R. (1986). *Time and the human interaction: Toward a social psychology of time.* New York: Guilford Press.

Miller, C.E., & Komorita, S.S. (1995). Reward allocation in task-performing groups. *Journal of Personality and Social Psychology, 69*(1), 80-90.

Miller, L.K., & Hamblin, R.L. (1963). Interdependence, differential rewarding, and productivity. *American Sociological Review, 27,* 768-778.

Mook, D.G. (1983). In defense of external validity. *American Psychologist, 38,* 379-387.

Moorman, R.H. (1991). Relationships between organizational justice and organizational citizenship behaviors: Do fairness perceptions influence employee citizenship? *Journal of Applied Psychology, 76,* 845-855.

Moreland, R.L., & Levine, J.M. (1982). Socialization in small groups: Temporal changes in individual-group relations. In L. Berkowitz (Ed.), *Advances in experimental social psychology* (vol. 11, pp. 137-192). New York: Academic Press.

Moreland, R.L., & Levine, J.M. (1988). Group dynamics over time: Development and socialization in small groups. In J.E. McGrath (Ed.), *The social psychology of time* (pp. 151-181). Newbury Park, CA: Sage.

Mullen, B., & Cooper, C. (1994). The relation between group cohesiveness and performance: An integration. *Psychological Bulletin, 115*(2), 210-227.

Naylor, J.C., Pritchard, R.D., & Ilgen, D.R. (1980). *A theory of behavior in organizations.* New York: Academic Press.

O'Dell, C. (1987). *People, performance, pay.* American Productivity Institute.

O'Dell, C. (1989, November). Team play, team pay—new ways of keeping score. *Across the Board,* 38-45.

Patten, T.H. (1977). *Pay: Employee compensation & incentive plans.* New York: The Free Press.

Porter, L.W., & Lawler, E.E. (1968). *Managerial attitudes and performance.* Homewood, IL: Richard Irwin.

Pruitt, D., & Drews, J. (1969). The effect of time pressure, time elapsed, and the opponent's concession rate on behavior in negotiation. *Journal of Experimental Social Psychology, 5,* 43-60.

Roy, D.F. (1955). Efficiency and 'the fix': Informal intergroup relations in a piecework machine shop. *American Journal of Sociology, 60,* 255-266.

Roy, D.F. (1960). Banana time: Job satisfaction and informal interaction. *Human Organization, 18,* 158-168.

Saaverdra, R., Earley, P.C., & Van Dyne, L. (1993). Complex interdependence in task-performing groups. *Journal of Applied Psychology, 78*(1), 61-72.

Seers, A., & Woodruff, S. (1997). Temporal pacing in task forces: Group development or deadline pressure? *Journal of Management, 23,* 169-187.

Shea, G.P., & Guzzo, R.A. (1987). Groups as human resources. *Research in Personnel and Human Resources Management, 5,* 323-356.

Shepperd, J.A. (1993). Productivity loss in performance groups: A motivation analysis. *Psychological Bulletin, 113,* 67-81.

Sundstrom, E., DeMeuse, K.P., & Futrell, D. (1990). Work teams: Applications and effectiveness. *American Psychologist, 45,* 120-133.

Terkel, S. (1972). *Working—People talk about what they do all day and how they feel about what they do.* New York: Pantheon Books.

Thompson, J.D. (1967). *Organizations in action.* New York: McGraw-Hill.

Tjosvold, D. (1984). *Cooperation theory and organizations. Human Relations, 37,* 743-767.

Tuckman, B.W. (1965). Developmental sequences in small-groups. *Psychological Bulletin, 63,* 384-399.

Tuckman, B.W., & Jensen, M.A.C. (1977). Stages of small group-development revisited. *Group and Organizational Studies, 2,* 419-427.

Van de Ven, A.H., Delbecq, A.L., & Koenig, R. (1976). Determinants of coordination modes within organizations. *American Sociological Review, 41,* 322-338.

Vroom, V.H. (1964). *Work and motivation.* New York: Wiley.

Wageman, R., & Baker, G. (1997). Incentives and cooperation: The joint effects of task and reward interdependence on group performance. *Journal of Organizational Behavior, 18,* 139-158.

Wagner, J.A., III (1995). Studies of individualism-collectivism: Effects on cooperation in groups. *Academy of Management Journal, 38,* 152-172.

Wagner, J.A., III, & Moch, M.K. (1986). Individualism-collectivism: Concept and measure. *Group and Organization Studies, 11,* 280-303.

Worchel, S.A., Coutant-Sassic, D. & Grossman, M. (1992). A developmental approach to group dynamics: A model and illustrative research. In S. Worchel, W. Wood, & J.A.Simpson, (Eds.), *Group process and productivity* (pp. 181-202). Newbury Park, CT: Sage.

Youndt, M.A., Snell, S.A., Dean, J.W., & Lepak, D.P. (1996). Human resource management, manufacturing strategy, and firm performance. *Academy of Management Journal, 39,* 8.

PERFORMANCE MANAGEMENT IN A FOURTH WAVE SOCIETY

SYSTEMIC MEASUREMENT, EVALUATION, AND INCENTIVES FOR DEVELOPING A HOLONIC PERFORMANCE MANAGEMENT SYSTEM

Robert E. Debold, Jr.

ABSTRACT

Society seems to be rearranging itself into what is commonly referred to as the post-capitalist society—one in which the means of production and human progress is being radically transformed. This transformation of society is creating a "knowledge wave." The idea of knowledge as opposed to information or hard assets will be acknowledged as the prime driver of organizational performance in the ensuing future.

If we look at any form of social organization in either the pre- or postcapitalist societies that have any degree of coherence and stability, we find they are hierarchically structured. The universal characteristic of hierarchies is the paradox of the "part" and the "whole." However, wholes and parts in an absolute sense just plainly do not exist, neither in living organisms nor in social organizations. What actually

Advances in Interdisciplinary Studies of Work Teams, Volume 6, pages 187-218.

ISBN: 0-7623-0655-6

exists are sub-wholes which simultaneously display characteristics of both wholes *and* parts. Managing performance in the knowledge-driven organization will require the understanding of "holarchy" (Koestler, 1989).

INTRODUCTION

The purpose of this chapter is to provide a structure that can be superimposed upon any organization by its leadership and management to help plan, develop, and maintain performance results in a knowledge-driven society. A novel, but well founded approach will be discussed and proposed that for organizations of individuals to achieve any form of productivity in the knowledge-based fourth wave society, principles of open hierarchical systems will be addressed. As part of this discussion a holonic performance feedback system, combined with the cybernetic principles surrounding the viable system model will be shown to be an archetype, which takes on a very basic and simplistic balancing loop structure.

At the 1964 World's Fair in Flushing Meadows, New York, a predominant theme was "progress is our most important product." The idea of progress had been around since the end of World War II, and it was to become the major focus for society and the business community for over two decades. The West, including the United States, was deeply into rebuilding the postwar economy up to the mid sixties, when another derivative social phrase, "do your own thing" emerged, putting a more individualized spin on the progress theme. Lovelock's (1987) seminal book on the Gaia hypothesis followed Carson's (1964) landmark contribution, *Silent Spring*, with the realization that we are all intricately interconnected in the living biological and social fabric of the planet. Lovelock was the catalyst needed to bring this realization into social/cultural awareness.

The dramatic visual in the *Whole Earth Catalogue* (1994) cemented our understanding—the wonderful blue and white orb from the moon just floating out there alone in space—we realized in this instant we just might be a "complete living system" and not simply moss growing on an inorganic globe.

A major paradigm shift occurred over a century ago in 1896, when American philosopher and educator, John Dewey (1859-1952), claimed that stimulus and response are not unidirectional, nor do they operate independently. He proposed that *responses influence stimuli.* A circuit or loop occurs back to the stimulus in the form of feedback to a goal-directed action. His proposition spawned the ideas surrounding control systems.

Control systems began being developed and were in wide use by the 1930s. They were specifically designed to replace human operators in jobs calling for the control of important variables in mechanical devices like the boilers in steam engines. Nevertheless, the fact that the behavior of such systems resembles the purposeful behavior of living organisms was almost completely ignored by behavioral scientists, including those working in field of organization science.

Along the way, many writers expounded on our inability to comprehend the problems and the deep woven connections between systems. It seemed obvious to many that a way to understand the structure of life was needed. Correspondingly, our metaphors were beginning to change from mechanistic to organic. The structural biology of Gaia was creeping into a variety of science disciplines. The monarch butterfly migrating from Mexico to Canada might have an effect on the weather in Idaho, and we set out to determine how that might occur.

The groundbreaking book, *Cybernetics* (Wiener, 1948) defined the emergence of control system concepts as expanding into human behavior. He defined a science of "effective communication and control in man and the machine" (http://www.gwu.edu/~asc). Beer later redefined cybernetics as "the science of effective organization" (http:www.gwu.edu/~asc). The amount of think time and energy consumed on the topic over the ensuing 50 years since Wiener's contribution has been enormous.

The traditional industries that have driven the postwar economy are being outrun by a whole new set of technologies: microelectronics, robotics, biotechnics, software systems, and telecommunications. A genuine global economy is emerging as financial markets have globalized and the major economies are converging investments and financial attention on rapidly developing new industries. The wealthiest individuals no longer own the land and the oil beneath it.

Our social organizations are also changing in structure (Drucker, 1993). Western individualistic capitalism is being challenged by communitarian capitalism in the fast-growing countries along the Pacific Rim. Maybe this attempt to put the group before the individual will outperform the traditional market mechanism and fare better than socialism. However, then again, perhaps a blending needs to occur.

The main purpose of this chapter is to address the dualism that is embodied in the problem of the individual and the group as it impacts our organizations and their ability to remain viable, that is, to perform to expectations, and to continually renew themselves. The structures of our organizations today are under attack, especially traditional hierarchy (Beer, 1981). The hierarchy is isomorphic to a "command and control" (C^2) form of management. Yet, while C^2 is under siege, it remains solidly intact, even as downsizing, and reengineering are played out. Teams have emerged as an answer to the empowerment thrust, but have yet to produce the desired performance results.

Rather than dismantling, I propose an evolving integration of the part and the whole, the individual and the group, to enact a more highly differentiated and inclusive organization form in order to move to the next level. A structure is needed that can be enfolded into any organization enabling the leadership and management to plan, develop, and maintain performance, while handling change and organization renewal. This structure will be shown to be holarchic in design, using concepts of control systems from cybernetics. The reification will need to be both philosophical and operational.

Toffler (1980) proposed an approach to analyzing social change which he called "wavefront" analysis, which serves as a foundation for this chapter:

> it looks at history as a succession of rolling waves of change and asks where the leading edge of each wave is carrying us. It focuses our attention not so much on the continuities of history (important as they are) as on the discontinuities--the innovations and breakpoints. It identifies key change patterns as they emerge, so that we can influence them. (p. 13)

For well over two hundred years, second wave societies have looked upon the organization and the individual as totally separate. This dualism—much like the mind-body dualism of Cartesian philosophy—the disassociation of the part from the whole, needs to be both philosophically and operationally integrated in order to take full advantage of the two approaches. This synthesis points to a structure that can be described as a holarchy. The universe that we know is structured holarchically, that is, like the Roman god Janus, having two faces looking in opposite directions. One face is turned toward the part, the subordinate level, and resembles an independent whole, while the face turned toward the superior level, looks like a dependent part. This particular structure is evident in all things in the universe. Wilber (1996) has graphically displayed this in his "four corners of the universe" theory.

Our organizations are increasingly complex. It has become nearly impossible in a C^2 structure to manage all the inputs and outputs in any organization and at the same time be expected to produce results which, for the most part, are beyond the capability of the organization. The problem of autonomy versus control has come to the forefront and is taking center stage. Yet if we look at any form of social organization in either the pre- or postcapitalist societies that have any degree of coherence and stability, we find they are hierarchically structured.

The universal characteristic of hierarchies is the existence of the paradox, part and whole. A part is considered a subassembly, a fragment, an incomplete element, which when considered by itself does not have a legitimate existence. A whole, on the other hand, is complete in itself and needs no further explanation. The complexities of this apparent oversimplification are actually deceptively robust. However, wholes and parts in an absolute sense just plainly do not exist, neither in living organisms nor in social organizations. What actually exists are sub-wholes which display simultaneously characteristics of both wholes *and* parts. Managing performance in the knowledge driven organization will require the understanding of "holarchy". A *holarchy* is an evolved concept of a hierarchy. Koestler (Koestler, 1989, p. 100) has coined the term *holon* to represent this object of duality.

THE FOURTH WAVE

The fourth wave is an idea discussed in a book by the same name (Maynard & Mehrtens, 1993). It is a logical yet derivative concept of Toffler's historical wave theory originally propounded in *The Third Wave* (1980). These waves are described categorically by Toffler as (1) agricultural, (2) industrial, and (3) information. Toffler says the waves are not as simplistic as the generic titles evoke. The designations only portray in a conceptual sense the patterns and forces that shape business, economies, politics, religions, global affairs, and social power interrelationships. Toffler's generic designations required three books (*Future Shock*, 1991a; *Third Wave*, 1980; & *Powershift*, 1991b) to best portray how these events have affected current life and might shape the future.

Before Maynard and Mehrtens (1993) contribution, one could extrapolate current social evolution to arrive at the theory that if Toffler's wave theory is truly accurate, and if each wave is behaving as a tighter and shorter bell curve relative to time, then the third wave might very well be cresting *now*. One would logically ask, just what might be taking its place in the same fashion causing the industrial age to give way to the information wave?

There are a number of seeming underlying patterns emerging from the current wave. One happens to surround the word *light*. George Bush, in his famous inaugural address of 1989, raised our collective consciousness when he called to task the creation of "1000 points of light" (http://www.bartleby.com/124/pres63.html). This is very much in line with contemporary authors and their development of the changes in society. If one takes a more expansive perspective of the term *light*, then we are talking about a respiritualization. This refers to the idea that people will see the need to get in touch with their inner selves. This is being abundantly played out in the desire to improve the quality of life not in the collection or acquiring of assets, but in the need to have more time to spend with families, in self-improvement and enjoyment of the arts (Rifkin, 1995).

The seeming overabundant problems evident in society today might be considered the "chaos" period as waves overlap and create undulating eddies almost appearing randomly. In the third world countries, the second wave is alive and kicking while the third wave is barely being understood in the nations (United States, Canada, Europe, Japan, etc.) that logically ought to be the home(s) of the third wave. Our accounting systems are sadly second wave based and take into account only hard smokestack assets on the balance sheets. It has been 500 years since Pacioli (*De Computis et Scripturis*, 1494, Venice) published his seminal work on accounting, and we have seen virtually no innovation in the practice of enterprise accounting—just more rules—none of which has changed the framework for measuring the third wave information society. Old paradigms die hard, even as the new wave is firmly cresting over our heads.

As third wave information industries advance to the extreme mechanization of the previous first and second agricultural and industrial waves, the fourth wave will be taking hold of the social fabric. This appears to be inevitable. Recall that in second wave affairs, the dominant relationships driving social governance were market focused, and they took precedence over traditional relationships; human worth was measured in commercial terms. At the beginning of the fourth wave, the market sector and government will play a diminishing role in day-to-day affairs. If the employed have more free time due to a shortened work week, while the unemployed have idle time due to lack of work, there seems to be two alternatives that would result: (1) increased social unrest and with it the problems of crime and homelessness, or (2) the evolving of the "third sector" which includes community activities, nonprofit organizations, volunteerism, or advocacy organizations. This is not a new concept. French social scientists have been using the term social economy to make the distinction between the third sector and a market-exchange economy. The social economy is measured in outputs, not salaries or revenue. A new accounting system will take shape as the social economy takes root.

However, are we missing an opportunity today in places like Russia or Ukraine? The fourth wave, if it truly is a bifurcation point, will be a chance for western market experts to jump to a social economy, or what is commonly labeled as "communitarian capitalism." If the experts are traditionalists, unfolding from the old and enfolding the new, creating a new holon will be quite difficult.

By taking into account how Toffler describes and generically defines the three major waves (he dismisses the hunter-gatherer wave), one gets a sense that human evolution can be viewed as playing a leading influencer role from which, and the degree to which, all other human events precipitate in lockstep. The emergence of the industrial second wave from the first wave, was not a necessary but a sufficient condition by which various derivative events such as the Marxist form of economics, development of management as a discipline to produce results, and the systematic development of crafts (which led to the development of organized labor called unions) evolved. Thus, the question becomes, what leading influencer is acting on our systems today that will become the next wave? Here is a short list to summarize a few of the events taking place right now that may be playing such a role:

1. Ubiquitous information
2. Emergence of the understanding of knowledge versus information
3. Emergence of systems thinking (General Systems Theory)
4. Ubiquitous communication
5. Emergence of cybernetics as a tool for all disciplines
6. Development of knowledge creation and storage technologies
7. Mass technological displacement in all industries both second and third wave
8. Regionalism encroaching on the nation-state and redefining sovereignty
9. Philosophy reemerging as an important discipline to be studied
10. Emergence of the need for organizational learning
11. The advent of virtual reality communication

12. The inability of all theologies to harmonize and reduce moral entropy in the face of the consequences of advanced technologies (e.g., euthanasia, abortion, DNA manipulation, cloning, etc.).

The fourth wave can be extrapolated from Toffler's analysis. It appears to be firmly grounded in Gaia-like concepts which are evolving out of the third wave society and will have the technological capability to integrate social results with indirect economic gains. The ideal fourth wave society has the potential to sharply reduce unemployment (Meyers, 1996; Rifkin, 1995); it will have the ability to reduce the stigma of underemployment (Rifkin, 1995); it will have the requisite variety to allow the remaining market-based economy to improve the quality of life; it will have the diversity to allow the government to mandate compulsory work, much like the U.S. mandated compulsory conscription to military service (pre-1982) and still mandated compulsory education, yet it will change the way we think and care about our neighbors—most people will begin to see the connectedness of *everything*. It will demand a new vision for education and life skills training (Adler, 1982). It will ask for increasing requirements to understand how our second and third wave organizations will perform in this new society. The fourth wave can be deemed the *knowledge wave*.

THE ORGANIZATIONAL PARADOX

Since the focus of this chapter is the performance of organizations, a brief look back at the foundations is necessary. It was Frederick Winslow Taylor (1911) who first applied knowledge to the study, analysis, and engineering of work. He didn't use the term *management*, even though it was the topic of his book, but rather *knowledge*. Taylor's main motivation was the creation of a society in which workers and owners could share a common interest in productivity and likewise build a harmonious relationship by applying knowledge to work. This began to occur as the second wave took over society. The rural landlords who once were dominant in the regions where they owned the land, gave way to new elites: corporate chieftains, media giants, and bureaucrats. This new power broker was the architect of mass production, distribution, communication, democracy, education, and consumption (Toffler, 1990b), the fourth wave society in some ways exemplifying Taylor's vision.

If the fourth wave is an emergence of knowledge, then it is important to understand the distinction between the third wave driver—information—and the evolution to knowledge. To understand the distinction between information and knowledge, a mapping of the evolution of cognition, or the way we organize our views of reality is useful. A starting point is the ontological view from the field of epistemology. Bellinger's (1995) linear holarchic view of the evolution of data →

information → knowledge → wisdom, provides a high level description of how this cognitive process develops.

The hierarchy begins with data. Data are just pointless (or meaningless) signals in space in time, without reference to either space or time. The key concept is *out of context*; data alone have no context or relationship with anything else. And, since they are without context, they exist without a meaningful relation to anything else. Bellinger (1995) states that when we recognize a datum of data, if it gets our attention at all, our first action is usually to attempt to find a way to attribute meaning to it. And, we attribute meaning by associating it to other things. This association creates a relationship. Thus, information can be described as an understanding of relationships between elements of data.

For example, a hydrogen atom is an element, a datum, but when related to two oxygen elements, then the information about water results from the relationships contained among the data in their makeup. In other words, when two elements of oxygen are associated to one of hydrogen, an understanding of the relationship between the data creates a thing called a "molecule of water."

> While information entails an understanding of relations between data, it generally does not provide a foundation for why the data is what it is, [*sic*] nor an indication as to how the data is [*sic*] likely to change over time. Information has a tendency to be relatively static in time, and linear in nature. Information is a relation between data and quite simply is what it is, with great dependence on context for its meaning, and with little implication for the future. (Bellinger, 1995, http://www.outsights.com/systems/kmgmt/kmgmt.htm)

Figure 1 represents how this holarchy unfolds. Information cannot be created without data, while knowledge emerges when patterns of information create context for the recipient (Drucker, 1993). Nonaka and Takeuchi (1995) claim that knowledge is context—specific and relational, and unlike information, is about beliefs, commitment, action, and meaning.

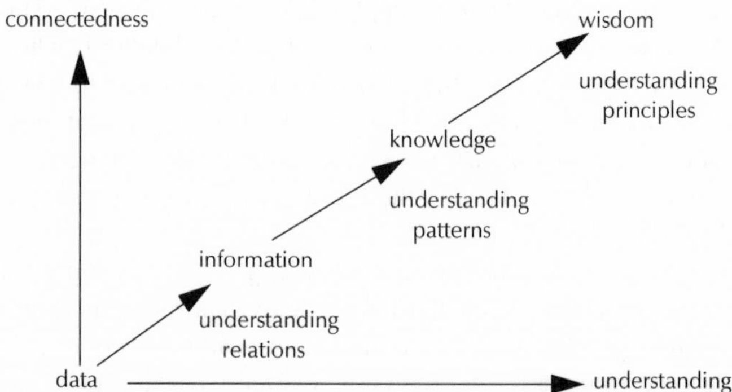

Figure 1. The linear view of the evolution of knowledge from data. Knowledge requires context and an understanding of patterns

Extending beyond relationship is pattern. Pattern contains within its definition the ability to be recognized as a model. As an ideal, pattern can allow one to develop ideas by which action can be taken. When relationships become fused into a pattern within a context, then and only then, does a basis for decisions to act in the future exist. When information that exists and becomes available is *associated* into a pattern, it allows the human (as opposed to an animal) to begin ordering and structuring the human experience. Associativity is the impression made on a thinking organism that many similar inputs can be collapsed on a prototype (model). Cognition—the act or faculty of coming to know—entails the ability to view a pattern from many facets and still recognize the pattern. Thus, pattern is a necessary but not sufficient condition for knowledge. One should only decide to act when a basis for knowledge exists.

Kuhn's (1962) ideas regarding the contextual development of patterns into paradigms, and Bronowski's (1978) belief concerning the internalizing of the patterns of the outside world parallel Nonaka and Takeuchi's (1995) description of information as a flow of messages, and knowledge, while created by that very flow of information, is anchored in the beliefs and commitment of the holder.

Nevertheless, if knowledge requires context and the understanding of the relationships between the information objects that make up a knowledge object, then a paradox exists in our current organizations. It is precisely the ability to create and retain knowledge to *learn* that the organization is unable to perform. Taylor's (1911) vision of using knowledge applied with scientific principles was lacking one fundamental principle not available to Taylor—the knowledge of feedback and control—which became available during the first half of the twentieth century in the form of general systems theory and the more human-oriented science of cybernetics. They were both rooted in the findings of the laws of thermodynamics.

Even though scientists could now calculate how machines could function with maximum efficiency, in the fast changing markets of the third wave society, Taylor's scientific principles began to be seen as a recipe for disaster. His process-oriented formulas had workers assigned to perform individual tasks. Taylor reduced processes to the smallest possible element. The idea of isolating parts of a task to control them was vintage second wave industrial methodology. It was clearly in line with reductionist philosophy that had been flourishing for well over five centuries

Taylor's process-oriented scientific management principles and the principles and practices of control theory never amalgamated. Thus, pursuit of greater efficiencies, that is, through improving products and manufacturing processes while our organizations struggled to adapt to fast moving changes, has become increasingly apparent. The reliance on mechanistic, reductionistic practices has resulted in great stresses to our organizations and human resources, specifically as the pace of business life has accelerated.

CHANGING DIMENSIONS OF PERFORMANCE

Performance practices during Taylor's era for the most part still exist today, and are found in the use of financial accounting. From the profit and loss (P&L) statement to the balance sheet, other indices such as return on investment (ROI) or return on net assets (RONA) are derived. The move away from pure rationalism to humanistic management created additional measures primarily focused around the individual.

The majority of new managerial ideas—like cross-functional teams, self-managed work groups, and networked organizations—are responses to the inadequacies of the models of Taylor and others. The individual performance measures, however, tend to be still reductionistic, because of measurement limitations.

Since organizations are structures set up according to a plan that are designed by some person, group, or class for the deliberate and express purpose of achieving certain goals (Abrahamsson, 1993), the conditions for their existence are important and should be considered in any model which attempts to define and predict performance. However the utilization of the organization as a resource itself is limited by the economic, technical, and political factors surrounding it. The individuals that make up the organization are themselves sources of feedback; motivation and behavior are significant and should be factored into any model.

Abrahamsson (1993) states that organizational theory has three major problem areas. The first is in determining how to make the organization efficient; the second is how to satisfy the goals of the mandator; and the third is how to offset the inevitable emergence of bureaucracy. The mandator is the individual or individuals who have taken the initiative to establish the organization and/or have raised funds to underwrite its existence. They carry more influence concerning goals than stakeholders and can have a different slant on organization effectiveness, in some cases compromising effectiveness and performance, and in other cases preventing their compromise.

These conditions of existence appear as the result of the evolved position on the lifecycle curve proposed by Adizes (1988). The mandator is paralleled as the founder in the Adizes model. In the early stages of an organization, the founder *is* the organization and personifies performance. Adizes describes bureaucracy as being the bellwether symptom for an organization that is no longer performing and, in fact, is in a rapid death decline. He describes radical prescriptions to correct an organization in this state. By the time bureaucracy sets in, performance has deteriorated so badly, it can hardly be measured.

Adizes'performance model is an organic model built on the logistic life-death curve all living organisms conform to. It is what we see in nature and is remarkably similar to the bell-shaped normal distribution curve also referred to as "Gaussian." Adizes says that to perform efficiently and effectively at any stage in the lifecycle, specific techniques of management must be present.

The view of performance along the lifecycle in Adizes' model requires feedback to determine what management technique would be appropriate. This idea of flexibility with control is eminently performed in complex adaptive systems. The complex adaptive system makes decisions in a decentralized fashion. Any number of decisions and decision-making processes occur on a local level and are maintained throughout according to some canon of rules. The operation fluctuates and controls itself according to external feedback; adjustments to the operation are made with respect to the responses to the environment.

Complex adaptive systems display the primary characteristics of cooperation, autonomy, and control. The notion of control is not from the C^2 sense, but rather the ability to manage input via feedback and maintain behavior based on a set of patterns (rules) according to thresholds.

Complex adaptive systems have several things in common. First they exhibit self-management or self-control along with autopoeisis (self-creation) and autonomy (self-regulation). Second, they are capable of learning through feedback from the environment by embedding experience in their actual structure. Third, they are constructed as open and hierarchical while providing for flexible specialization. The best complex adaptive system model to make patterns is the human brain. The brain is continually reconfiguring the connections between neurons and dendrites in response to external and internal stimuli. It learns through feedback, and notwithstanding its triune construction (reptilian, neo-mammalian, mammalian), is completely modularized in areas that specialize in specific functions.

Thus, in order to avoid the problem of bureaucracy, an alternate structure is one that allows subagents to network information paths; operate as self-managed but organized; respond to feedback and adjust behavior to the environment; learn from experience; embed the experience into the structure; and reap the advantages of specialization without being stuck in rigidity. The above qualities are found and closely match organic, complex, and adaptive systems.

CONTROL

The term *control* tends to be construed pejoratively. The idea that humans can be controlled has become a paradox in itself. Davidson (1996) states that people don't obey orders without much consideration anymore. The assumptions on which the C^2 philosophy rests are unable to aid performance. Davidson states:

> The Vietnam experience reinforced the trend [C^2]. People simply stopped taking for granted that what they were told to do was the best thing to do, either for themselves or their organizations. Because of the new world of unprecedented competitive intensity and accelerating change, this would eventually prove to be a source of great strength, but in the short-term it dealt a death blow to existing methods of management. "Manage" comes from an Italian word that means, literally, to train or handle horses. Today's workforce simply won't be treated that way. (p. 169)

The metafluctuation between the middle of the sixties and the beginning of the seventies (Jantsch, 1980) broke down all barriers regarding the notion of control. The protests were mostly against the restrictions of life and became powerful processes which have helped shape the urge to find new structures. Individuals no more want to be controlled than wild mustangs. However, our organizations are still in the business of control. While the storm did blow over by the eighties, the mental structures had changed—the world has not been the same since.

The seeds that had been planted in understanding control systems and cybernetics led directly to the concept of autopoeisis—the ability of a living organism to continuously renew itself and to regulate the process in such a way that it retains its structural integrity. Whereas a machine is geared to producing a specific prod-

		Commuting Time in Minutes for Train A									
	55	90	100	70	55	75	120	65	70	100	
	75	95	75	110	65	85	110	65	85	80	
	65	60	75	65	95	65	65	90	60	65	
	80	60	65	60	70	65	85	90	65	60	
	80	55	65	60	70	65	70	60	75	80	3730
x(bar)	**71**	**72**	**76**	**73**	**71**	**71**	**90**	**74**	**71**	**77**	746
X(dblbar)											**74.6**
R	25	40	35	50	40	20	55	30	25	40	
R(bar)											**36**
UCLx(bar)	=	95.48									
LCLx(bar)	=	53.72									
p(bar)	0.02			100.3	=UCL for the runs						
				48.95	=LCL for the runs						

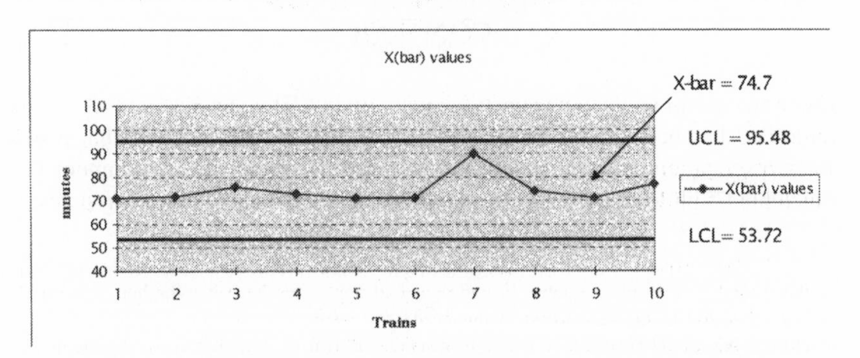

Figure 2. Commuting times show a system in control within upper and lower control limits

uct, performance can be considered static, but a biological cell is primarily concerned with renewing itself *as it produces and achieves its short and long-term goals* (Jantsch, 1980). This creates, as Jantsch points out, openness at the top, in the same way that Koestler (1989) described the performance of a holarchy.

The cybernetic and systems meaning of control concerns the ability to maintain statistical averages (means). A system in control (Deming, 1993) is one that, over an observable period of time, will maintain its performance within upper and lower control limits. Managing systems without knowledge of how the system is performing over time is what Deming calls tampering.

Figure 2 is an example of an x-bar chart, which is used to monitor processes with continuous data. The x-bar chart reflects how the process average varies over time. The example shown displays a sample of how many minutes ten trains take to arrive at a station. The example in Figure 2 is a process that is *stable* or *in statistical control*. The process is in control because there are no points outside the control limits.

Tampering has to do with the aspect of taking action on a stable system (one in statistical control) in response to fixing a faulty output. Acting on a lack of knowledge about the statistical control of a system has a high probability (near 100%) of achieving exactly the opposite of what is intended. By tracing a stable system upstream (Deming, 1993), management gets leverage on sources of faults and mistakes. This involves the controlled reduction of common causes in the system.

COOPERATION AND AUTONOMY

The disassociation between cooperation and autonomy has been of paramount significance in the management of organizations. How can one be autonomous and cooperate at the same time? This fundamental problem is one of the primary reasons for the concepts and introduction of holonic manufacturing systems (HMS) in the manufacturing industry. In living organisms, the self organizing, autopoietic tendencies of holons exhibit the theoretical foundation for guiding the cooperation among holons, leading to a globally near-optimal performance. One of the measuring methodologies in HMS, is the use of Langrangian relaxation techniques of mathematical optimization (Luh & Hoitomt, 1993).

In HMS modeling, the interactions among cells (holons) are being performed without intruding upon the individual subunits' private information and decision autonomy. The modeling uses object-oriented technologies (C++ and Smalltalk programming languages) that encapsulate data and methods within objects (autonomy), and at higher object levels can clearly delineate the responsibilities and relationships (cooperation).

Likewise, holonic performance management system (HPMS) architectures can be modeled by using whole-part relationships. Contrary to conventional performance management systems, an HPMS is managed in a distributed manner by

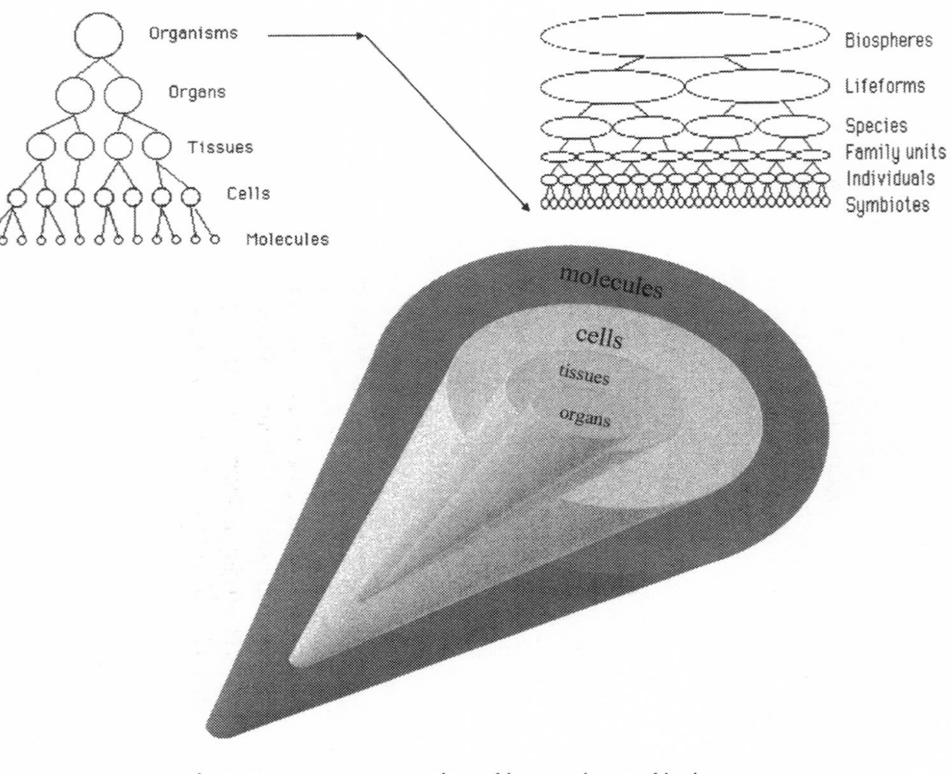

Figure 3. Two examples of hierarchies of holans.
Each holon at its particular level displays cooperation and autonomy

empowering the autonomy of the system elements or holons. Individual holons define their activities based on their local knowledge, and decide their behaviors by cooperation with other holons through their standardized feedbacks. An HPMS would therefore possess the following properties.

Modularity

A holon (element) has its local information, decision-making autonomy, and standardized interfaces. It consists of a physical component and information component. The physical component corresponds to physical entities such as individuals, products or machines, while the information component possesses information, decision-making capabilities, and mechanisms for managing the physical component.

Decision Autonomy

Decision authority and responsibility are distributed among individual holons throughout the system. Holons are self-sufficient, and possess the capability to create and control the execution of their own plans and/or strategies.

Cooperativeness

Individual holons are not expected to operate with absolute autonomy. Rather, they collaborate with other holons to decide their activities, function within the system constraints, and adjust their behavior according to the coordination of feedback.

Recursivness

An HPMS consists of holons at various levels; holons at different levels share homologous structure. For example, similar knowledge interfaces, language definitions, and message protocols are needed to ensure cooperation among holons within a level and across levels.

Figure 3 is a very primitive, but basic picture of two hierarchies of holons. A holon is a node in a holarchy. A holon looks up for what it needs to cooperate with and integrate with. It looks sideways for what it needs to associate and gain input from (in some cases it may compete). It looks down for what it evolved from and wants to command-by-exception. Each holon cannot be fully explained by or predicted by a study of its parts. It is something more. A holon is also part of something bigger than it is being affected by. However, at the same time it has a high degree of autonomy—it has a life of its own.

Prior to the introduction of cybernetics as a potential management tool, the conventional wisdom for managing cooperation and autonomy was to demand coop-

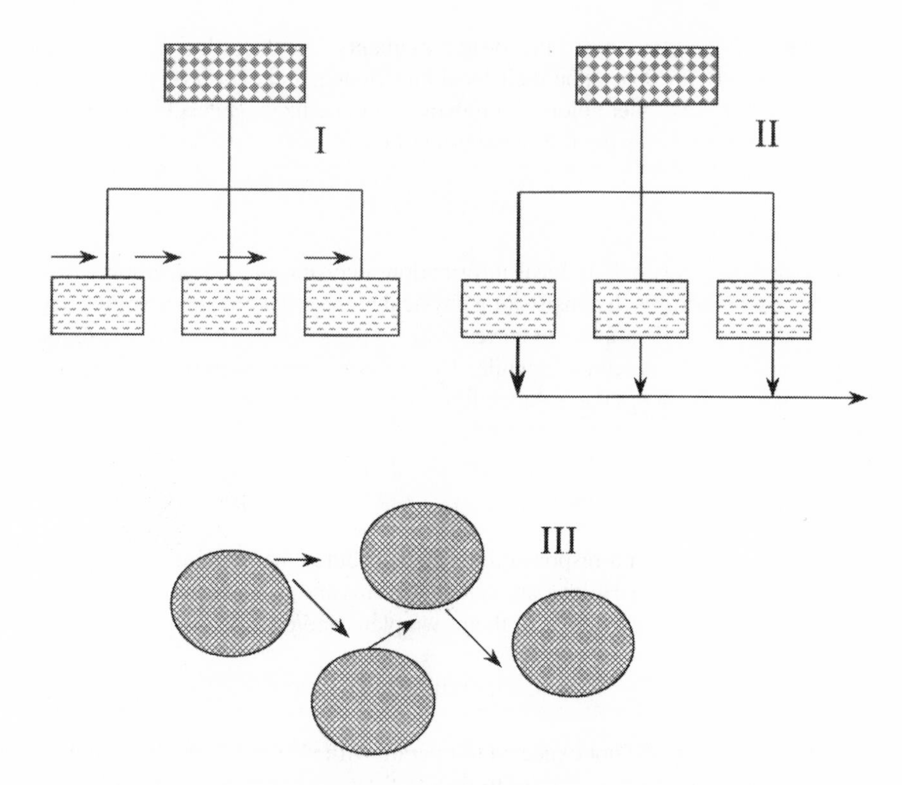

Figure 4. The Cartesian-mechanistic models of feedback/control systems. Type three exhibits random feedback and connections and is the least structured.

eration and limit autonomy. This is just not practical, as we have been building the argument for both to exist in parallel.

Hersey and Blanchard (1982), proposed a way of handling the dichotomy by coordinating the task structure against the follower's capacity to perform a task. They have defined the need for three systems of control. In Type I, the boss controls the activities of the subordinates hierarchically and horizontally as if in a production line environment. In Type II, the boss still controls the activities but task functions are combined. In Type III, which is the least structured, the group members act as separate decision makers and functional units, having the same advantages as the managers in Types I and II, that of direct customer contact. Figure 4 portrays these three types of control systems.

Using the Cartesian methodology as seen in Hersey and Blanchard (1982), the problems of complexity, rate of change, and interdependency cannot be met effectively. One can see that the limitation of non-holonic thinking creates and exacer-

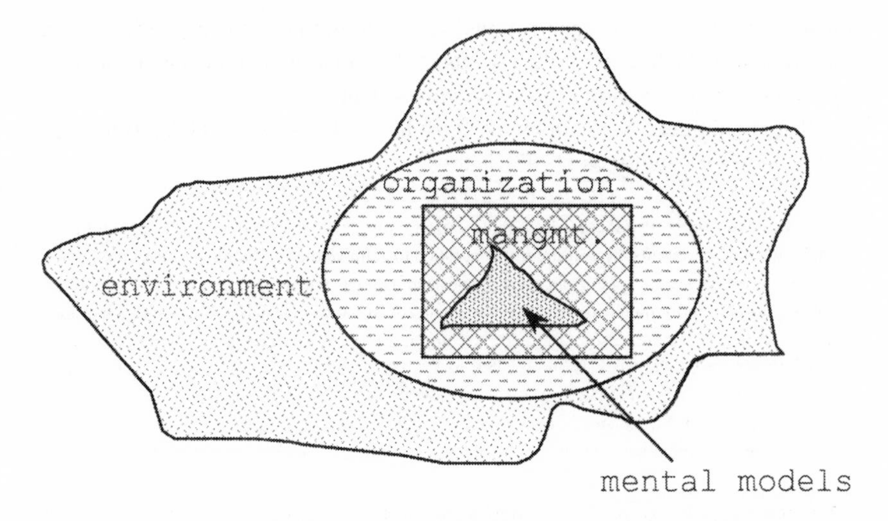

Figure 5. Embedding of models, managers, and operations within the environment. The holonic structure is the essence of the viable model.

bates problems inherent in the mechanistic, hierarchical approach. The requirements of the late twentieth century of meeting the needs of customers continually, learning and adapting rapidly, and maintaining the learning and adaptation to improve effectiveness must be met. What is needed is a model, a methodology for building the model, and a discipline for providing a general framework.

Clemson (1991) discusses this particular problem of autonomy versus control in a model that incorporates the ability of Beer's (1981) viable system with the law of requisite variety. The law of requisite variety states that the level of regulation of a system is achievable in the degree to which the complexity of the regulator is greater than or equal to the complexity of the system. Clemson builds a holonic four-tiered operational unit that includes the environment with recursively embedded operations, management, and mental models, respectively. Figure 5 is adapted from Clemson's ideas about operational units in organizations showing the embedded holons.

According to Clemson (1991), due to the embeddedness and incompleteness of each holon, management at best can only partially control the operation, and likewise, the operation can at best partially control the environment. This implies and imputes the need for *self-regulation.* A manager of a system is unable to completely control any system he/she/they are embedded in, simply because they lack the variety to do a complete job. By variety we mean requisite variety as defined above. It is the ability of the regulator to achieve control over any situation. Clemson suggests most large organizations clearly do not have the capacity to match isomorphically the totality of the aspects of the situation that management is concerned with. Figure 5 shows this to be self-evident.

Self-regulation can be again seen against the backdrop of the Gaian rules found within evolution. Sahtouris (1997) states:

> It is abundantly clear that the needs and interests of individual cells, their organ "communities" and the whole body must be continually negotiated to achieve their dynamic equilibrium, commonly called balance. Cancer is an example of what happens when this balance is lost, with the proliferation of individual cells outweighing the needs of the whole. In the same sense a mature ecosystem—say a rainforest—is a complex ongoing process of negotiations among species and between individual species and the self-regulating whole comprised by the various micro and macro species along with air, water, rocks, sunshine, magnetic fields, etc. As Soros recognizes: "Species and their environment are interactive, and one species serves as part of the environment for the others. There is a feedback mechanism..." among levels. (p. 6)

Nevertheless a group of organizational units needs someone or something to "cohere" it, that is, provide it the infrastructure to maintain a separate existence, its viability, its sovereignty, while performing its function and purpose within the larger enterprise. In this context, changes within holons could create negative antagonisms that would need to be resolved via negotiations. This must be accomplished by the next higher-level holon. Thus, it is imperative to maintain what the cyberneticians call homeostasis, that is, all critical variables within the system maintained within normal limits to ensure the organization functions effectively.

An aid to understanding is the organizational chart that indicates how each part of the business formally relates to the other parts. However, the organizational chart lacks the dimension of control: how will input be optimally converted to output? How will information be transformed into knowledge? In the fourth wave organization, control has to do with the creation and dissemination of knowledge of an extent and complexity that is beyond the capacity of the boxes (management) to absorb. Thus, the boxes found on traditional organizational charts are not coherent wholes, that is, they are not designed to enable the requirements of autonomous entities.

A parallel of the viable holonic model can be made once again when considering the Gaian viable model:

> The evolution of living systems, as well as their ongoing livelihood, is an improvisational dance of negotiations among individual parts and levels of organization—among the holons in a holarchy. This dance is energized by the self-interest of every part and level, choreographed by compromises made in the tacit knowledge that no level may be sacrificed without killing the whole. At its best it becomes elegant, harmonious, beautiful in its dynamics of non-antagonistic counterpoint and resolution. This, I believe, is the Popper/Soros vision of the Open Society, where the interests of all levels would be open to discussion and thereby harmonized. (Sahtouris, http://www.radical.org/LifeWeb/articles/globalize.html)

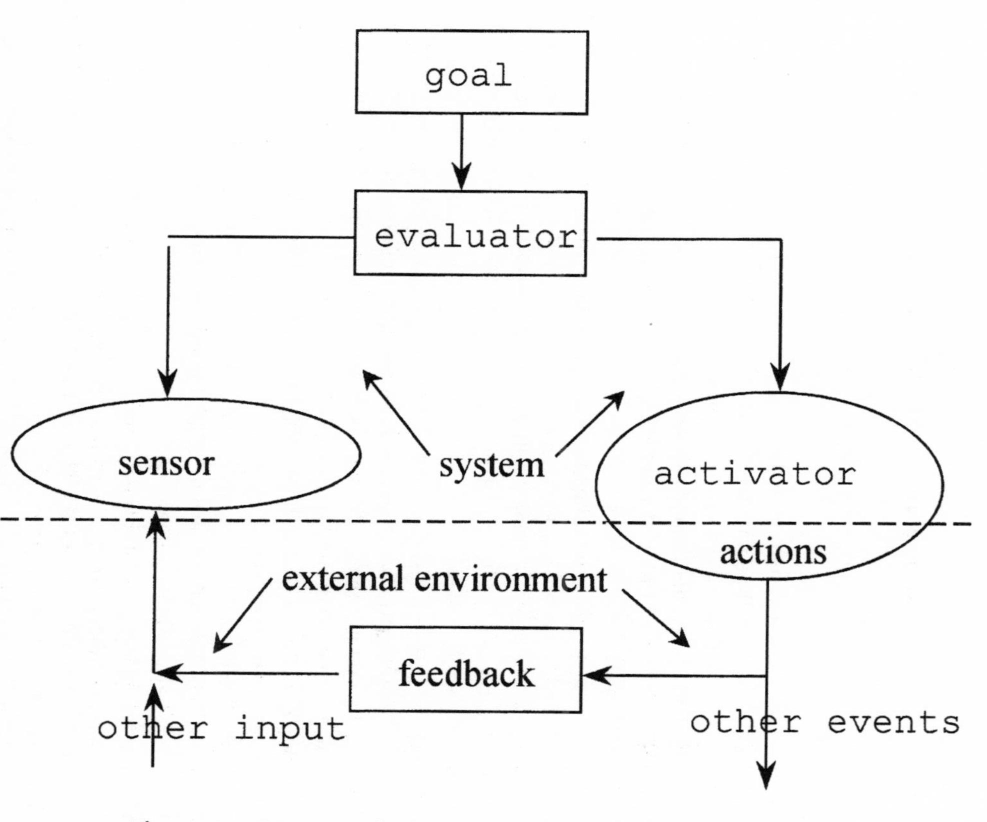

Figure 6. Diagram of cruise control negative feedback system

Thus, the structural inadequacy in the traditional model of the organization is clear: a new model is required which will actually work as a model. However, the organizational chart was not designed to model the aspects of the enterprise we wish most to understand—the areas that have to do with *control*. Beer (1981) said it is as though we poured paraffin into a plastic model of a jet airliner in the hope that it would fly!

HOLONIC PERFORMANCE

Feedback Loops

Feedback provides a system with the ability to be self-regulated. The principle of self-regulation is a fundamental fact of all non-pathological hierarchies. In order to operate within some form of fixed rules, information from the environment must be made available to the system (the organism) based on its progress toward some goal. This progress can only be evaluated against some known and previously set capability.

When the feedback becomes known, the controlling element, or regulator, must constantly adjust the course of the operation. Cruise control on an automobile is an excellent example of how this feedback operates (Cziko, 1995). Feedback is generally defined as a coupling of output to input. The current application is known as *cybernetics*.

There are many examples of control systems found in electronic devices today, but the most commonly used today, is as noted, automobile cruise control. It is designed to maintain a steady driving speed with no assistance from the driver. After the driver sets the desired speed, the unit uses this setting as the desired goal to achieve and maintain. Accordingly, the unit will increase or decrease the fuel input to maintain homeostasis of velocity. This is a negative feedback control system. The feedback from the output (actions on the environment) is used as input to sense and then compare against the goal. Figure 6 provides a systemic diagram of what is actually happening here.

The interesting thing about this particular control system is that the cruise control only uses feedback from one source: the car's speed (technically, from the intake manifold and the first plug). Thus, for emphasis, the term control is used as Beer (1981) and Clemson (1991) describe, that is, for maintaining the critical variables within normal limits to function effectively and maintain homeostasis, exactly like holistic management and holonic-cybernetic systems are designed to do. A control system is designed to control what it senses, not what it does. *Thus it controls its input, and not its output*. However, feedback without a hierarchy is like a grin without a cat (Koestler, 1989). The cybernetic approach demands that the feedback from the environment merely guide, correct, or stabilize the organism according to some preexisting patterns of behavior. Feedback doesn't alter

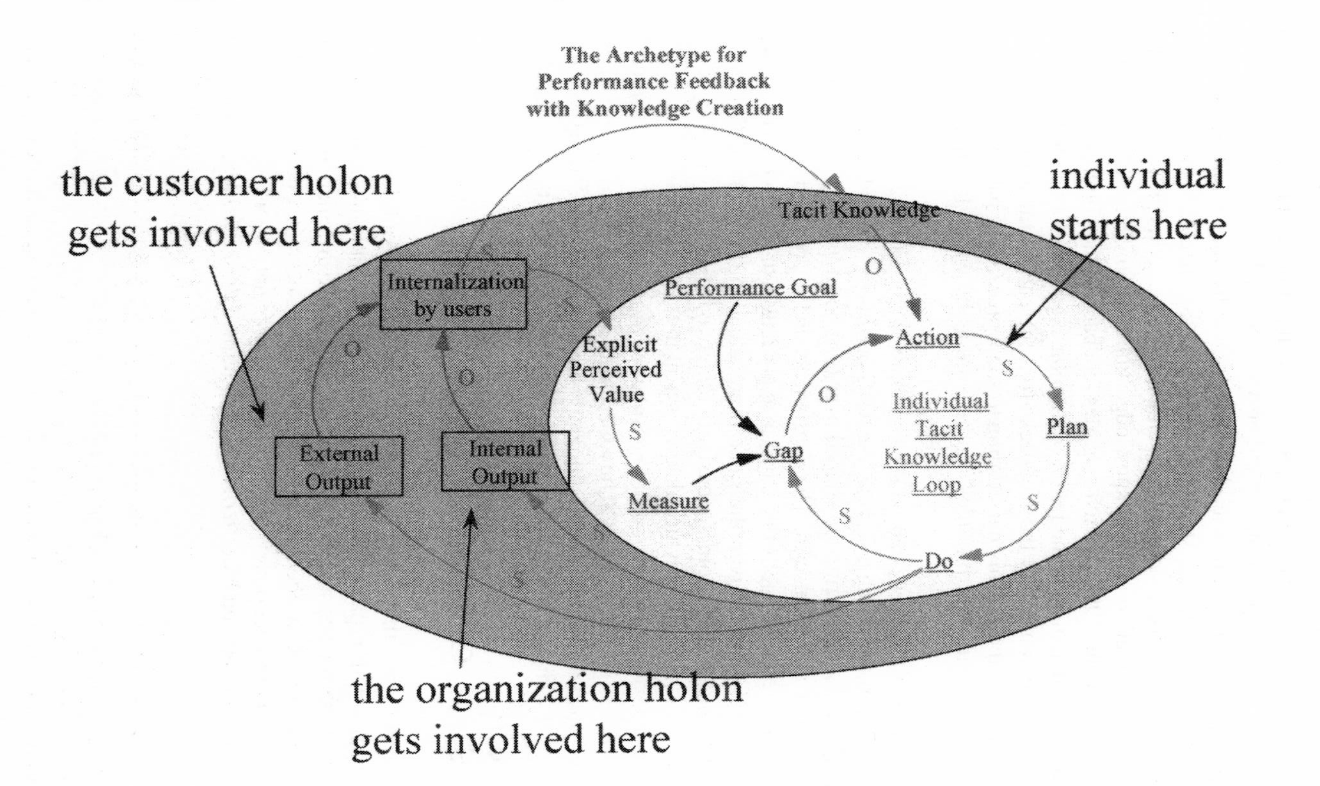

Figure 7. Holonic performance management plan-do-check-act negative feedback loops. Incentives flow as feedback from the act loop

the intrinsic patterns of the rule or process already existent and decided on by the organism. Figure 7 shows how an organization can mimic the negative feedback control loop to monitor performance and provide feedback of the output.

The interior holon in Figure 7 might be either an individual or an organization, while the enveloping holonic structure could represent its customers to which it provides value (the "do" module). The value chain (or circuit) provides for internal and external creation of tacit knowledge via the knowledge creation spiral (Nonaka & Takeuchi, 1995), which provides feedback to the holon for recalibration when mapped against some initial performance goal. Once again, the cybernetic approach demands that the feedback from the environment merely guide, correct or stabilize the organism according to some preexisting patterns of behavior. A holonic architecture provides for the patterns by which learning can take place.

This pattern development can be seen in many ways. For example, after one learns the patterns for riding a bicycle, the ways one keeps from falling over while pedaling forward and steering are shaped over time and somehow "burned in" to our neurons that move muscles. The constant feedback provided to one's eyes, touch and balance, merely tells one when to apply brakes or to turn to avoid a dangerous hole in the road. Given the same exact situation, two individuals would perform differently, albeit homologously. In fact, two individuals would have different levels of threshold by which danger is sent to the consciousness.

The higher-level system of conscious thought must interrupt the seemingly autonomic patterns of bicycling. Thus the changes in the environment, or the alterations of the external conditions, do not *cause* a process to occur, but modify the process via a higher level system (holon) in the hierarchy of the organism.

OPEN HIERARCHICAL SYSTEMS

It is abundantly clear that nature is constructed and unfolds-enfolds as an organization of parts-within-parts. All living matter and all stable inorganic systems have a parts-within-parts structure. This structure leads to articulation, coherence, and stability, and where the structure is not obvious, the mind provides it. For example, we can project butterflies in inkblots and camels in clouds. This structure is open at the top and Koestler (1989) has termed these structures *open hierarchical systems*. All living organisms and systems are thus open hierarchical systems.

When we look at holons, we observe that holons have dual tendencies. First a holon attempts to preserve its identity and then assert it. It seems to act as a semi-autonomous whole. For example, in the body, the heart has a unique identity that can be preserved even if it is transplanted, while performing its functions in relation to the entire body. Second, a holon performs as an integrated part of a larger whole. This duality of behavior is inherent in all hierarchic structures, is a universal characteristic of life. Wilber (1996) has described the evolutional process as a coherent one:

> All evolutionary and developmental patterns proceed by holarchization, by a process of increasing orders of wholeness and inclusion, which is a type of *ranking* by *holistic* capacity. This is why the basic principle of holism is holarchy: the higher deeper dimension provides a principle, or a "glue", or a pattern, that unites and links otherwise separate and conflicting and isolated parts into a coherent unity, a space in which separate parts can recognize a common wholeness and thus escape the fate of being merely a part, merely a fragment.

Thus, open hierarchical systems are essentially rule-governed in their behavior and characterized by enormous flexibility and freedom of choice. These flexible strategies are guided by *feedbacks*. Living systems, also known as complex adaptive systems, strive to maintain homeostasis. Homeostasis is the maintenance of critical variables within normal limits so that an overall organism can continue to function effectively (Clemson, 1991). Practically, this means that the balance between stability and the rate of change must be within a statistical probability that would consider the system under control. That is, x-bar has been calculated and the values of actual performance are within the calculated 3-sigma plus and minus upper and lower control values.

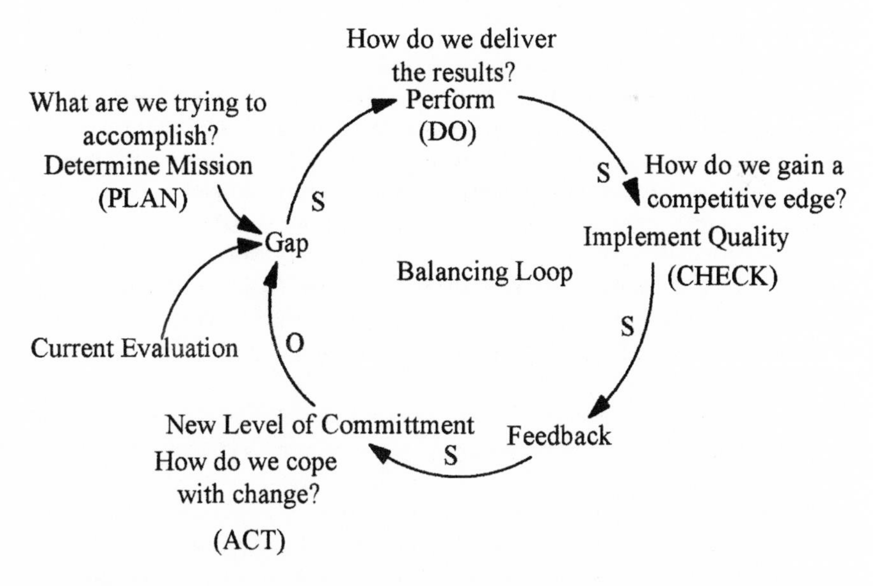

Figure 8. Balancing loops produce feedback structures that help to maintain homeostasis and maintain a balance goal attainment in accordance with grand strategy

Figure 8 is another systems diagram showing the balancing (negative feedback) loop by which an organization maintains homeostasis while improving on quality. The "check" stage provides the gap analysis by which the balance between stability and the rate of change is within a statistical probability to be considered under control. The value of a holon's performance can be evaluated with what Deming (1993) calls profound knowledge. Ackoff (1974) has a very similar definition of performance, that is, the relationship between controlled variables and uncontrolled variables (common and special causes), which includes the limits within which each of the controlled variables can be manipulated.

Semi-Autonomous Sub-Wholes

Take the living system most intimately familiar to all of us: the human body. We have long known that our bodies behave as a community of cells. It has a central nervous government that continually monitors all its parts and functions, ever making intelligent decisions that serve the interest of the whole enterprise, and an immune defense system to protect its integrity and health against unfamiliar intruders.

More recently, microbiology has revealed the relative autonomy of individual cells in exquisite detail: every cell constantly making its own decisions regarding what to filter in and out through its membrane, and which segments of DNA to retrieve and copy from its nuclear gene library for use in maintaining its cellular welfare (Dawkins, 1989).

It is clear that the needs and interests of individual cells, their organ communities, and the whole body must be continually negotiated to achieve systemic and dynamic equilibrium—balance. Cancer is an example of what happens when this balance is lost, with the proliferation of individual cells outweighing the needs of the whole. In the same sense, a mature ecosystem—for example a rainforest—is a complex ongoing process of negotiations among species and between individual species and the self-regulating whole comprised by the various micro and macro species along with air, water, rocks, sunshine, magnetic fields, and so forth. "Species and their environment are interactive, and one species serves as part of the environment for the others. There is a feedback mechanism…." (Sahtouris, 1997, p. 6) among levels.

In social hierarchies, we see the same structure and behavior. Looking inward toward an individual, we see a person is made of organ systems, organs, cells, molecules, and atoms. An outward view would falsely give one the impression that he or she is a complete and unattached whole. However, an individual is really a smaller part of various larger social units. We can begin at the level of families, clans, tribes, and move to neighborhoods, communities, counties, and cities. Other examples include government, academic, and business hierarchies. These hierarchies are termed open due to their infinite expansion moving out to the universe. The holonic cell displays this interrelationship with its environment by using the permeability of its cell membrane-boundary to exchange energy, waste and entropy with the larger holonic environment it is a constituent of.

Organization as Organism

Obviously metaphors have their limits including body models. However, bodies paradoxically trumpet mechanical metaphors of perfect societies running like well-oiled machines. They are something we all have in common regardless of our worldviews, our political or spiritual persuasions, and they do exemplify the main features and principles of all healthy living systems or holons, be they single cells, bodies, families, communities, ecosystems, nations, or the whole world. By understanding them we can assess the health of any particular living system and see where it may be dysfunctional. This, in turn, will give us clues to making the organization or enterprise healthier.

According to Sahtouris and others, some of the features of living systems (organisms) are:

1. Self-creation (autopoeisis)
2. Self-regulation (autonomics)
3. Self-maintenance
4. Self-reflexivity—(autognosis)
5. Embeddedness in larger holons
6. Input/output flow of matter/energy/information
7. Transformation—of matter/energy/information
8. Complexity—diversity of parts
9. Communications among parts (chemical, electrical, etc.)
10. Coordination of parts and functions
11. Balance of interests among parts, with whole and with embedding holons
12. Reciprocity of parts in mutual contribution and assistance (win/win economics)
13. Full employment of functional parts
14. Conservation of what works well
15. Creative change of what does not work well. (Koestler, 1989, p. 144)

Many of the above features represent goals of our organizations. We want them to constantly renew themselves, have the ability to learn, and the capability to adapt quickly to environmental changes. Holonic performance then requires the ability to know what is actually happening, determine capabilities and forecast the potentialities. There is no room for rigid, static, and top-down driven targets. The ability to adapt to sudden and rapid change requires an inherent built-in intelligence so that one unit can communicate easily with other units. This type of performance must operate in an autonomous, distributed, and cooperative system.

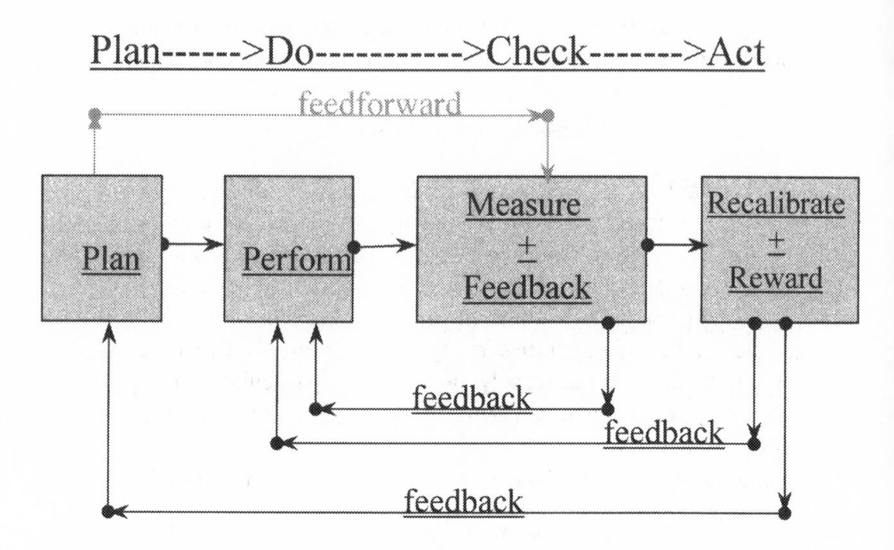

Figure 9. Shewart control loop providing feedback and feed-forward loops. Appropriate rewards and incentives help maintain a system "in control."

Holonic Performance Management

Holonic performance management systems (HPMS) are systems by which characteristics of open hierarchical systems are combined with Beer's (1981) viable system model from the field of cybernetics to produce appropriate feedback loops that will allow an organization to maintain homeostasis according to a balanced scorecard (Kaplan & Norton, 1996). Shewart's control cycle is used to promote overall circularity and provide the framework for logical feedback structures (Deming, 1993). Figure 9 is a generalization of how this feedback and control cycle should occur.

Central to the notion of HPMS is the understanding of the holon in the structure of complex adaptive systems as well as universal evolution (Wilber, 1996). Each performing unit is considered a holon, must cooperate with other holons at its level—from planning and scheduling to physical production—to the construction and delivering of knowledge. It is a new paradigm in measuring performance which challenges the C^2 structure of business and organizations. The many technical challenges will be commensurate with the pressure involved in restructuring the organization to accommodate the paradigm hurdle.

In an HPMS, rigid, static, and hierarchical organizations give way to structures that are more adaptable to change. The measuring systems must accommodate this change. The units must be designed so that they have inherent intelligence

Table 1. Comparison of Second/Third Wave Performance Topology Compared with Fourth Wave Holonic Feedback Concepts

Second and third wave topology	*Fourth wave (holonic) topology*
Strategies of the global organization communicated down as tasks	Strategies of the global organization communicated down as sub-goals
Master-slave relationships	Customer-Supplier relationships
Deterministic communication patterns with fixed message structures	Universal signal scheme with varying communication patterns
Labor replaced by automation in an isomorphic fashion	Automation enhancing the knowledge capabilities of labor and complementing the efforts
Focus on financial historical measures	Focused on balanced ratios of real-time actuality to capability to potentiality
Producing high volume/low variability	Producing mass-customization (high-low volume/high variability)
Mechanistic application	Organic application
Newtonian-reductionist metaphor	Complex-adaptive metaphor

(Beer, 1981) with the ability to communicate with other units in a real time fashion.

Organization

By giving an organization autonomy at the appropriate holonic level, a structure can be transformed from any hierarchy to a systemic structure that displays autonomy and cooperation in appropriate doses. By doing so, the sub-units will have:

1. Elimination of the need for central command and control
2. Self configuring capabilities
3. Reconfiguring flexibilities
4. Capability to integrate human work to grand strategy
5. Reusability anywhere in the organization
6. Capability to realize individuals' role in the organization's success.

It should be reemphasized that holons are not clones. Each holon has an individual uniqueness even though it is evolved out of homologous patterns as other holons at its particular level. The most important facet however, is that the desirable holonic behavior, especially that of individuals, is incorporated into the overall structure. Feedback loops must be developed to provide information by which the ratio of actual to capable performance can be determined. The entire range of activities that will be performed must be surfaced and provided real-time performance ratios.

A primary task in designing a holonically structured organization is to develop the organizational mappings such that the man-machine interfaces are cooperative. On one hand, the individuals cooperate to achieve grand strategy goals. On the other hand, the organizational holons are distributed to obtain and use computing power, requisite geographical space, and communication networks

Applying the principles of complex adaptive, open hierarchical systems, as a design rule for performance feedback metrics, will lead to new ideas of organizational control loops using technology that will be compatible with holonic or viable system organizational structures. Thus, an effective HPMS is composed of holons, each of which contain people, communications network, methods for feedback, and a physical processing system.

Grand Strategy

The primary feature then, of HPM, is the installation of communication channels in both the vertical and horizontal direction. It is the horizontal-level communication that is the most troublesome to implement. The vertical transmission can take the structure as defined in the viable systems model and progress from a top-down to bottom-up to a hypertext organization which functions as middle-up-down (Nonaka & Takeuchi, 1995). The within-species communication model is an appropriate analogy to display horizontal communication. This is why humans must be given feedback from the technology they interact with to enable them to provide autonomous decisions within a pre-given decision range. Thus, an HPMS is an anthropocentric model and contains within its essence a revolution in the way we empower and control.

Davidson (1996) states that grand strategy is:

> about the transformation of management itself. Its [*sic*] purpose is to find a superior way of managing—the ultimate competitive advantage, and the only competitive advantage that can survive the kind of unpredictable and accelerating change of the era in which we live. (p. 2)

Davidson (1996) also says:

> Grand strategy thus ensures that we not only win today by beating our current competitors, but also that we go on winning tomorrow by beating evolution itself. It finally brings to the world of organizations what man has long struggled to bring to nature—the ability to manage change instead of change managing us...grand strategy deals with the connections between that competition and the performance of these resources. (p. 10)

Thus, in Davidson's view, the ability to align internal organizational processes with their environment, so that the organization can implement an external strategy, implies the challenge of performance. Thus, performance becomes how we deliver results and how we maintain homeostasis to reach potentials. This view aligns with the PDCA-holonic-cybernetic feedback loop seen in Figure 8. This archetype provides for the capabilities that must be measured: *homeostasis*—the ability to survive; *process efficiencies*—the ability to align internal resources; and *capability improvement*—the ability to learn from current performance and adjust the inputs to meet the expected outputs. Grand strategy encompasses the totality of achieving and maintaining organizational competitive health.

Davidson's (1996) implementation of grand strategy is concerned with performance of the organization. He wants our organizations to think strategically by bringing thinking to bear on the structure of the organization's relationships, to thereby encompass all those who have a stake in the organization and its performance.

Grand strategy linked to holonic concepts presents a decentralization concept where executive authority is essential to the effective management of the communication channels. HPMS remedies the problem that humans and automation are not perceived as being able to be integrated. Compatibility will be merged in terms of competence, skills, and decision abilities related to the retrieval of knowledge. The feedback systems, people, and the physical processing system form a subunit holon.

MAKING OUR ORGANIZATIONS WORK

"Far reaching changes are on the way" (Diebold, 1982, p. 106). These words spoken by Diebold in the Trueman Wood Lectures delivered to the Royal Society of Arts were almost understated. Diebold said that despite the great promise of modern technology, there existed a dismaying gap between what business can achieve and what we do achieve.

Diebold was referring to the increasing gap between the growth of the gross national product and the growth of knowledge. Between 1960 and 1980, the average compounded growth of the GNP has been about five percent a year for all governments and societies across the globe. This means that if the pace continues, a person who lives to be 80 will see the GNP grow to 32 times what it was when he or she was born. Diebold predicted that the limits to growth would slow and the poorer two thirds (the third world) would get the farmed-out, low level industrial production not desired by the wealthy one third. This eventuality has arrived. Diebold (1982) called this the multinational corporations' revolution (p. 107). He noted that knowledge was increasing at an annual rate of about five percent also. Using this percentage, in year 2060, about 97 percent of the knowledge existing in the world will be totally new knowledge. As Diebold (1982) states:

> However, I do not think that the social institutions we have created are geared to move forward at the same pace of advance as is now being achieved in the technosphere. Thus, I think that anyone who plays any part in these institutions—of which capitalist business is one—needs to recognize that the proper definition of a reasonably responsible behavior pattern is and should be constantly changing. (p. 108)

The makeup of American management has gone through three major shifts (Davidson, 1996; Drucker, 1993). In some degree, corporate leadership has evolved through waves of engineering-oriented, sales-oriented, and finance-oriented CEOs. This fits respectively with people such as Henry Ford/Thomas Edison, Thomas Watson/Alfred P. Sloan, and efficiency accountants, George Fisher/John Scully.

The problem of efficiency accounting has been exacerbated recently by the downsizing phenomenon. It is slowly giving way to a change-oriented focus where the primary concern is renewal and the response of promoting innovation. What exacerbates this situation, is the rate of change of human accomplishments. If one plots an envelope curve of anything from roughly 1800 to present, the tangent is almost a vertical line. Human capacity to travel at speed beginning with the horse and progressing to the train, the car, the propeller aircraft, jet aircraft, chemical rockets, and finally nuclear rockets, is an example of this rate of change. More recently, the memory capacity on computer chips is behaving in the same fashion.

The entire structure of society has a problem of adaptation. Concerning management or the organization, the same problem exists, maybe even to a greater degree. What is needed is a total reappraisal of our way of managing organizations. This entails a complete reconsideration of the traditional (industrial) organization. The problem of organizational adaptation is a problem of organizational planning, and should not be looked at as being solved by conventional wisdom. How to successfully run companies, or how to effectively organize and run government is no longer clear.

Beer (1981), Adizes (1992) and Davidson (1996) all have said that management is supposed to be the instrument of *change*. In the traditional paradigm, the manager is responsible to provide *control*. This means that we must design control systems and not just implement technology or the latest human resources fad. Planning processes must be improved in context with technological change. The question that needs addressing concerns how the organization should be managed, given that technology is rapidly changing and there are no easy answers regarding the control problem.

Holonic performance management system development is an attempt to provide a living systems analogy to augment the findings of organizational science. It provides a way of thinking about the organism and organization isomorphically. It incorporates cybernetic concepts with an understanding of the laws of large numbers. The viable systems model, being a holonic structure itself, provides a framework by which the concepts of autonomy and cooperation can be mapped from holonic characteristics.

Aristotle coined the term *entelechy*, which describes the realization of some function in which a potentiality has become an actuality. Organizations should be striving to achieve their potentials. In addition, in order to achieve entelechy, we must be able to measure performance. The model to measure performance has been defined in this chapter as operations under open hierarchical and holarchical principles. These principles provide an emerging structure to detect, recognize, measure, and adapt to changes in the internal and external environment of an organism/organization. These changes produce relative gaps in the unrealized potential of an organization's achievements, and when functionally organized, are able to detect these shifts immediately—to act fast—to perform. Holonic systems are viable, and all viable systems are autopoetic. If found to be absolutely true, organizations structured with holonic performance concepts supporting viable systems will be lasting and able to handle any exponential growth curve they encounter.

REFERENCES

Abrahamsson, B. (1993). *The logic of organizations*. (A. Forlag & W. Forlag, Trans.). Thousand Oaks, CA: Sage Publications.

Ackoff, R.L. (1974). *Redesigning the future*. New York: Wiley.

Adizes, I. (1988) *Corporate lifecycles: How and why corporations grow and die and what to do about it*. Englewood Cliffs, NJ: Prentice Hall.

Adizes, I. (1992) *Mastering change the power of mutual trust and respect*. Santa Monica, CA: Adizes Institute Publications.

Adler, M.J. (1982). *The paideia proposal*. New York: Macmillan Publishing Company.

Beer, S. (1981). *Brain of the firm* (2nd ed.). Great Britain: The Pitman Press.

Bellinger, G. (1995). *Knowledge management*. http://www.outsights.com/systems/kmgmt. kmgmt.htm

Brand, S., & Warshall, P. (1986). *Whole earth catalog*. Garden City, NY: Doubleday.

Bronowski, J. (1978). *The origins of knowledge and imagination*. New York: Oxford University Press.

Clemson, B. (1991). *Cybernetics a new management tool* (2nd ed.). Philadelphia: Gordon and Breach Science Publishers.

Cziko, G. (1995). *Without miracles universal selection theory and the second darwinian revolution*. Cambridge, MA: MIT.

Dawkins, R. (1989). *The selfish gene*. Oxford, England: Oxford University Press.

Davidson, M. (1996). *The transformation of management*. Boston: Butterworth-Heinemann.

Deming, W.E. (1993). *The new economics for industry, government, education*. Cambridge, MA: MIT Center for Advanced Engineering Study.

Diebold, J. (1982). *The role of business in society*. New York: AMACOM.

Drucker, P.F. (1993). *Post-capitalist society*. New York: HarperCollins Publishers, Inc.

Hersey, P., & Blanchard, K. (1982). *Management of organizational behavior Utilizing Human Resources*. Englewood Cliffs, NJ: Prentice-Hall.

Jantsch, E. (1980). *The self organizing universe: Scientific and human implications of the emerging paradigm of evolution*. Oxford, England: Pergamon.

Kaplan, R. & Norton, D. (1996). *The balanced scorecard*. Cambridge, MA: Harvard Business School Press.

Koestler, A. (1989). *The ghost in the machine.* (4th ed.) London: Penguin Group.

Kuhn, T.S. (1962). *The structure of scientific revolutions*. Chicago: The University of Chicago Press.

Lovelock, J.E. (1987). *Gaia*. New York: Oxford University Press.

Luh, P.B., & Hoitomt, D.J. (1993). Scheduling of manufacturing systems using the Lagrangian Relaxation Technique. *IEEE Transactions Automatic Control, 38*(7), 1066-1080.

Maynard, H., & Mehrtens, S. (1993). *The fourth wave*. San Francisco: Berrett-Koehler.

Myers, P.S. (1996). *Knowledge management and organizational design*. Newton, MA: Butterworth-Heinemann.

Nonaka, I., & Takeuchi, H. (1995). *The knowledge creating company*. New York: Oxford University Press.

Rifkin, J. (1995). *The end of work*. New York: Tarcher-Putnam.

Sahtouris, E. (1997). *The biology of globalization.* http://www.radical.org/LifeWeb/Articles/globalize.html.

Taylor,F. W. (1911). *The principles of scientific management*. New York: Harper and Brothers.

Toffler, A. (1980). *The third wave*. New York: Bantam.

Toffler, A. (1991a). *Futureshock*. New York: Bantam.

Toffler, A. (1991b). *Powershift*. New York: Bantam.

Wiener, N. (1948). *Cybernetics: Or control and communication in the animal and the machine*. New York: Wiley.

Wilber, K. (1996). *A brief history of everything*. Boston: Shambhala Publications.

COMPLEX SYSTEMS AND SENSEMAKING TEAMS
CONFLICT, CONNECTEDNESS, AND LEADERSHIP

Dennis Duchon, Donde P. Ashmos, and Maria Nathan

ABSTRACT

This chapter applies a complex adaptive systems view of organizations to the use of teams. Such a view suggests that teams have a new role essential to the organization's well-being, that of sensemaker. Sensemaking teams will likely have to rely on behaviors traditionally considered inappropriate. For example, viewing organizations as complex adaptive systems in which sensemaking is an ongoing activity that allows collective understanding to emerge through self-organization leads to understanding that conflict is an ever present phenomenon which is an opportunity to be embraced, not a problem to be defused. Moreover, it will be necessary for sensemaking teams to develop a collective mind which is a kind of connectedness different from the notion of cohesiveness that has been viewed as an ideal group practice. Finally, using teams for sensemaking will create new and different demands on the role of leader, demands which traditional views of management have not entirely anticipated.

Advances in Interdisciplinary Studies of Work Teams, Volume 6, pages 219-238.

ISBN: 0-7623-0655-6

INTRODUCTION

Traditional approaches to management have been based on the idea that the world is knowable. The world is knowable because, in the Newtonian view, it is a kind of mechanical system in which identifiable forces and fundamental laws of motion are in operation (Capra, 1982; Stacey, 1995). Such a system may be difficult to comprehend, but, armed with a knowledge of cause and effect, a manager could, over time, come to understand the forces and the laws. This knowledge along with good business practice would allow a manager to achieve predictability, order, and control in the organization. In recent years, good business practice has included the use of teams for problem solving and decision making. Teams, if properly used, have been seen as a mechanism for improving information processing and ultimately helping the manager achieve order.

However, all of the traditional managerial practices, including the use of teams (Ashmos, 1997), require rethinking. There is an emerging view of organizations which, in contrast to the traditional view, sees the universe as an indivisible, dynamic whole whose parts are essentially interrelated and can be understood only as patterns of a cosmic process (Capra, 1982, 1996). One important assumption is that the future behavior of a system cannot be accurately predicted from its present state (Rae, 1986). Small changes in initial conditions can sometimes produce large changes in outcomes. Events do not unfold with average regularity and adjustments rarely produce the desired effect. Extrapolating from historical data is thus neither feasible nor desirable. This emerging view of organizations confounds traditional thinking and suggests different ways of managing. The purpose of this chapter is to examine the role and uses of teams as suggested by the complex adaptive systems view of organizations.

ORGANIZATIONS AS COMPLEX ADAPTIVE SYSTEMS

The complex adaptive systems view is rooted in chaos theory (Prigogine & Stengers, 1984) and quantum theory (Herbert, 1985; Youngblood, 1997) and has provided significant insights into the design and management of organizations (Stacey, 1992; Wheatley, 1992). One of the significant changes a complex systems view of organizations offers is in reference to the role of control, particularly the idea of manager as controller. Traditionally a manager achieved order and control as a consequence of both knowing the "rules" and systematically applying those rules. According to a complex systems view, however, trying too hard to predict and control the environment is not practical, because the world is, essentially, unknowable (Rohrlich, 1987), and because we cannot separate ourselves from our observations of the world. The shift away from controlling things is possible, because in the complex systems view it is not an understanding of the properties of a thing that are as important as the *connections* or *relationships* between things. Each thing derives its meaning, not from its inherent properties, but rather from the relationships it has with other things. Nothing exists independent of its relationship with its environment (Herbert, 1985). Relationships, moreover, possess a nonlinear quality (Prigogine & Stengers, 1984), such that the past does not necessarily predict the future.

If organizations are systems of coordinated actions among individuals and groups whose preferences, information, interests, and knowledge differ (March & Simon, 1993), then nonlinearity is a more plausible assumption than linearity. Nonlinearity exists, because the differences in preferences and knowledge mean that organizations possess the capability of making many different responses, and thus ensuring many different outcomes. Through multiple interactions and connections organizations have the potential for many responses, yet traditionally they are studied as if they were simple, linear systems operating in simple environments guaranteed to produce a single, predictable outcome. Complexity is about the condition of the universe which is integrated yet too rich and varied to be understood in simple, mechanistic, or linear ways.

Nonlinear, however, does not imply nonsystematic. For example, McDaniel (1997) notes that while the future is unpredictable, and many responses are possible, the behavior of organizations is not random. There is order in systems because of their emergent properties (Cohen & Stewart, 1994) and self-organizing capabilities (Prigogine & Stengers, 1984). The richness of interactions among the parts and between the system and its environment allows the system as a whole to undergo spontaneous self-organization. That is, it can adapt a form and processes which best suit a current need (Wheatley, 1992).

Youngblood (1997) contends that complex, self-organizing systems possess the following characteristics: (1) they are "wholes" beyond the sum of their parts; (2) they are open systems operating far from equilibrium that continually renew themselves through reciprocal interactions with their environment; (3) they have the capacity to spontaneously create new forms, order, behavior, and structure; (4) they are interconnected in complex, nonlinear ways; and (5) they are both autonomous from and merged with their environments.

Overman (1996) uses the metaphor of a river to illustrate how a complex, self-organizing system might operate. Looking at a swiftly flowing river one can see both the inevitable smooth flow of water and within that, points of turbulence and instability. It is always the same river, yet never the same water that flows by. The river is constantly in the process of self-organization as some structures such as channels and eddies continuously form and dissipate. The river's flow is irreversible, but within any segment there are complex fluid dynamics that can affect the flow far downstream. A river may periodically overrun its banks and form new channels in sudden floods, or it may meander slowly across acres of land over long periods of time (Overman, 1996)

Applying Youngblood's (1997) complex systems characteristics to the river one can see that the river itself is a "whole" beyond the sum of its water molecules. The river is a system far from equilibrium continually renewing itself as water flows through it. The river's flow and turbulence continually create new forms and patterns for itself, and the river's fluid dynamics describe many nonlinear connections. The river as an entity is separate and distinct from its environment, yet at the same time it is an important feature of the landscape and, as the river changes, so does the look and character of the landscape itself change.

The swiftly flowing river as an example of a complex system also assumes that the river is "far from equilibrium." Far from equilibrium implies something about the inherent dynamism of a system. Although assumptions of dynamism do not on the surface seem particularly troublesome, consider that such assumptions are important because they stand in contrast to traditional economic and management theories which assume that a state of equilibrium and balance is a system's desired end-state. Complex systems are believed to survive and thrive *because* they are not "in balance." Only when systems are forced out of balance will they experiment and explore their space of possibilities. Such exploration helps the system discover new patterns of relationships, different structures, and innovative ways of working (Prigogine & Stegners, 1984). While a system cannot sustain being far from equilibrium all the time, and the goal of experimentation is to find a way to restore some kind of equilibrium, the system can sustain periods out of equilibrium or even sustain portions of itself being out of equilibrium more or less constantly. Without being far from equilibrium the system stagnates and dies. Thus, the complex system is always seeking its balance, but absolute equilibrium is not one of its goals.

Although perfect equilibrium may not be one of its goals, the complex system is making a continual effort to find some kind of balance, and balance can emerge through a process termed "self-organizing." Self-organizing occurs when connections and interactions among system members produce coherent behavior, even in the absence of a hierarchy which traditionally has been assumed necessary to ensure order. Self-organizing can thus be seen as an effort to find "balance." The patterns of coherent behavior are spontaneous, that is, not decreed or designed by any individual. Such spontaneous patterns may disappear, or they may have a long lasting effect, that is, self-organizing may produce a temporary balance or it may produce a long-term balance. In the latter case an evolution has occurred.

In complex systems terms evolutions occur at bifurcation points. Bifurcation points are choice points (Overman, 1996). Prigogine and Stegners (1984) note that at a bifurcation point a system can make three choices. It can allow something outside itself to make a choice, or it can follow its own deterministic rules to make a choice. The third possibility is that the system itself makes a situation-dependent choice, and it is the action taken to carry out a situation-dependent choice (i.e., creating a new coherent pattern of behavior) which constitutes self-organizing. Note that spontaneous self-organizing is not a far-fetched idea in human social systems. Informal groups of managers in a business, for example, form networks and no central authority organizes them. Yet these informal networks behave in a controlled way and are often critical to the organization's well-being (Stacey, 1992).

Self-organizing is about developing new, coherent patterns of behavior, and patterns of coherent behavior are the result of organizational actors trying to make sense of their circumstances (McDaniel, 1997). As the environment changes, people inside organizations must constantly reassess not only their own roles but also the relationships that exist between the organization and the environment. Thus the process of sensemaking is an important element of a complex system, because the system must continually interpret and make sense of its environment if it is to survive (Weick & Roberts, 1993). Sensemaking is the first step for the system not simply to find balance, but to find balance in a new, more adaptive form.

Sensemaking is not about "understanding" in order to make rational decisions, although making effective decisions is an important outcome of sensemaking. Sensemaking is about an effort to stabilize the environment by paying attention to it (Weick, 1995). That is, sensemaking is not so much about imposing order as it is recognizing an emergent order. Sensemaking is not about "finding" predictability, as it is about "making" predictability in order to come to grips with an unpredictable world (McDaniel, 1997).

By adopting a complex systems view of organizations in which sensemaking takes on heightened importance, it is easy to see that teams play an expanded role from the role played in the traditional organization. Teams will continue to have the traditional evaluation/decision-making role, but they will also have a new role as a catalyst in that the team may be the best place, the best mechanism, for adaptive ideas to emerge. The resources and connections available to teams allow them a chance to create more and better new (adaptive) ideas than can an individual. New ideas can be chosen to replace old ideas, and the new ideas chosen can then set the organization onto new paths, new behaviors, some of which can lead to successful adaptations to the organization's environment.

Applying Youngblood's self-organizing characteristics, it is possible to see the team is an organized unit, a whole, which can function beyond the sum of its individual members. The team members each bring in data from the environment, and so the team itself can be seen to engage in reciprocal interactions with the environment. The team can create for itself new patterns, new behaviors, new structures. Not only are the team members interconnected, but the team itself is interconnected to the larger system in many, complex ways. The team functions as both an autonomous unit and as part of a larger environment with which it is merged.

Teams will also have an expanded role because of their capacity to acquire and process data about both the system of which they are a part and the larger environment to which the system itself is trying to respond. This is the role of "sensemaker." That is, the team, not scattered individuals and not larger agglomerations of individuals, may be the appropriate unit of action for "making sense." Then, as sensemaking is seen to be an important function of teams, three aspects of team behavior will require rethinking: conflict surfacing, connectedness, and leadership.

A NEW ROLE FOR TEAMS: SENSEMAKING

Teams are a special type of task group, consisting of two or more individuals responsible for the achievement of a goal or objective, and also possessing a common commitment (Katzenbach & Smith, 1993). There are two perspectives that have generally guided the consideration of using teams: a psychological perspective and a cognitive perspective (Ashmos, 1997). The psychological perspective promotes the use of teams as a motivating mechanism for minimizing worker resistance to change, and for increasing worker acceptance of team outcomes and commitment to goals and decisions (Locke & Schweiger, 1979). The cognitive perspective proposes the participation of workers in teams as a way to reduce uncertainty by increasing the information, knowledge and expertise that is brought to bear on a particular problem or task (Locke & Schweiger, 1979). From the cognitive perspective team members may contribute to problem solving, or they may improve their understanding of decisions they are supposed to implement, by being involved in discussing and deciding those issues. Thus, either uncertainty about a problem solution or uncertainty about solution implementation is believed to be reduced by the use of teams. Both of these perspectives, the psychological one and the cognitive one, make the same assumptions: that managers can improve their ability to better match intentions with outcomes through the use of teams, and that the ability to control organizational functioning is improved through the use of teams.

A major requirement of organizations, as important as the more traditional tasks of problem solving, decision making, and task accomplishment, is sensemaking. Sensemaking is an exercise in contextual rationality, an ongoing accomplishment that emerges from efforts to create order and make sense of what occurs (Weick, 1993). Recall that situation-dependent choice making is a characteristic of a complex system's self-organizing. Sensemaking, then, is a key means by which the system processes the ambiguity and equivocality (i.e., the data) that it finds, so that it may self-organize into new, more adaptive patterns. Sense is made of what is happening around and within the system (Weick, 1995). From a complex adaptive systems view, sensemaking is a prerequisite to successful adaptation because only by "making sense" at a bifurcation point can the system make a choice which proves adaptive and thus ensures its long-term viability.

When organizations engage in sensemaking they are attempting to answer the question, how do I know what I think until I see what I say? (Weick, 1995). Teams are often microcosms for the organization, and as such the team can be the mechanism for collectively seeing what it is the organization thinks. Teams offer the opportunity for an organization to test out ideas, assumptions, and interpretations of events, to argue and question, to develop shared meaning about the events and world it is facing. From this perspective, the purpose of teams is not to find answers to questions or to accumulate information for uncertainty reduction, rather it is to figure out what the questions to be asked are, to sort through the multiple cues and interpretations that create confusion and ambiguity in organizations.

The complexity of organizations makes it nearly impossible for any individual to be the "sensemaker" and teams are a way the organization can engage in collective sensemaking (Ashmos & Nathan, 1998). Sensemaking has seven properties (Weick, 1995) which can be the behavioral guides for teams engaged in sensemaking. First of all, when teams are used for sensemaking they help the organization in identity construction. That is, the team helps the organization discover what it "thinks" and "knows" about itself and its environment. Identity construction is the basis for imparting meaning to information inside an organization and, eventually, determining what problems need solving.

The second property of sensemaking involves retrospection. Retrospective sensemaking is an examination of past practices in order to learn (and unlearn) things about the current context (Weick, 1995). The multiple points of view afforded in teams makes retrospection a richer and more detailed process. Teams can also participate in the third property of sensemaking: the enactment process. What is commonly thought of as the "environment" is not so much a series of time/space events as it is a confluence of human efforts and actions aimed at creating meaning (Smircich & Stubbart, 1985). That is, there is no objective environment out there separate from one's interpretation of it (Burrell & Morgan, 1979). Thus, the organization creates (enacts) parts of its environment through selective attention and interpretation. Teams, representing the complex relationships in the organization, play an active role in this process.

The fourth characteristic of sensemaking is that it is a social process (Weick, 1995). A social process requires the participation of more than one person, and teams, with a diverse composition of people, can thus provide a rich setting in which sensemaking can occur. The lens of complex adaptive systems emphasizes the importance of connections, creating links and bonds through social interaction. The density of the connections is an important feature of both teams and complex organizations.

Weick (1995) describes sensemaking as ongoing (characteristic number five), suggesting that the task is never finished. Thus, teams as sensemakers, will be an ongoing phenomenon in organizations, a dynamic layer that gets added to whatever structure exists. The makeup and focus of teams will emerge in an ongoing fashion.

Sensemaking characteristic six has to do with extracting cues, and it is important that multiple collections of individuals be part of this process. Multiple perceptual frames that are represented in a team offer a better chance of achieving a workable reality for the organization. Finally, teams provide a forum for plausible storytelling (characteristic seven). Sense is often found in the context of a story which captures a plausible explanation for the environment.

These seven properties of sensemaking provide a context for understanding how team behavior may look different when the focus is sensemaking and the setting is complex adaptive systems. We propose that at least three dimensions of team behavior require closer examination: conflict surfacing, connectedness, and leadership.

SENSEMAKING TEAMS AND CONFLICT SURFACING

Traditional models of organizations stress the use of teams for task accomplishment and problem solving in order to help the organization return to a state of equilibrium. Restoring order, correcting deviations, and reestablishing stability are often the unstated goals of teams in traditional organizations. Total quality management programs, for example, promote teams as a primary mechanism for restoring order in internal processes so that more predictable and higher quality responses are made possible. In complex adaptive systems teams can purposely lead to just the opposite: disorder, deviations, and instability. That is, teams can operate, for a time at least, far from equilibrium and thus create for themselves opportunities for situation-dependent choices that might lead to new, adaptive patterns of behavior.

So, while teams in traditional organizations are encouraged to resolve conflict in order to achieve team goals, in complex adaptive systems teams can use conflict to move toward new collective understandings of the organization and its world. Conflict, despite the psychological stress and social anxiety it typically causes, can be seen therefore as a necessary part of sensemaking.

Yet using conflict to "make the organization better" (i.e., more adaptive) is an idea not likely to receive an enthusiastic reception, because harmony is one of the traditional goals of the team experience (Robbins & Finley, 1995).

The visible presence of conflict creates difficulties for most organizations because of a common aversion that people have to conflict, as noted by Maslow:

> We are inculcated with anticonflict values from childhood, and, as a result, most of us grow up with mores that sanction unquestioned authority. Disagreement is considered unacceptable; all conflicts are bad....We live in a society that has been built on anticonflict values. (Maslow, 1965, pp. 337-338)

Thus, conflict is traditionally seen as dysfunctional and something to be avoided.

Such a negative view of conflict, however, is problematic for organizations operating in turbulent worlds. The traditional hope of team harmony is lost in complex organizations because of the diverse nature of systems, with all parts of the systems connected to every other part.

From the traditional view of organizations, effective team behavior involves identifying the causes of conflict and removing them (Robbins & Finley, 1995) because they are viewed as barriers to team problem solving and goal achievement. From the traditional perspective, teams function best when conflict is minimized. Thus, much attention in the literature on teams and group development has focused on conflict resolution. While conflict resolution skills are important for people in organizations to have, perhaps even more important in contemporary organizations will be the need to learn to live with conflict and to let it help the organization to learn and grow (Tjosvold & Tjosvold, 1995).

Viewing organizations as complex adaptive systems in which sensemaking is an ongoing activity that allows collective understanding to emerge through self-organization, leads to the understanding of conflict as an ever present phenomenon. When teams are charged with helping manage the unknowable, conflict occurs, because multiple actors with multiple perceptual frames create multiple understandings of what is happening to the organization. For sensemaking, it is the surfacing of conflicting perspectives that is essential for getting to some new collective understanding. Teams charged with sensemaking are teams in conflict. To take the traditional perspective and attempt to eliminate the sources of conflict is to take away the team's ability to collectively make its way through the ambiguity and equivocality it faces.

In order to improve the way individuals work in teams, it is essential that a new perceptual frame be used that alters expectations about team functioning. First, it is important to understand why in complex adaptive systems teams will experience and create conflict. Second, it is also important to understand ways of counteracting our tendencies to think that conflict is something that is bad and is to be avoided.

Systems Characteristics That Lead To Conflict

First, teams are holograms. The concept of holograms is based on the notion that any part of the whole contains within it the "blueprint" for the larger whole itself (Bohm, 1983). Teams are a part of the whole organization. Some argue that the world is made up of wholes within wholes. Teams reflect the organization as a whole. Whatever values, conflicts, concerns, or paradoxes that exist in the organization, so too will they also exist in teams. Teams reflect the whole from which they are drawn and because of this will reflect the conflicts that exist in the whole. Similarly, we suggest that team members and especially team leaders must become the masters of conflict, letting it inform us, challenge our frames, and push us to new understandings that were not possible without the conflict.

A second reason why conflict is inevitable is that everything in the organization is interconnected and interdependent. A fundamental principle of complex adaptive systems is that everything is connected, that nothing is separate from anything else (Gagne, 1995). The reductionist thinking that pervaded traditional views of organizations led to separation of the parts of organizations which in turn led to the need for elaborate coordination mechanisms. In such a traditional environment teams are usually established for a specific purpose, are sometimes separated from much of the ongoing day-to-day functioning of the organization, and impermeable boundaries between the team and other parts of the organization appear. In contrast, complex systems theory proposes organizations as a complex web of relationships where everything affects everything else. When one organizational member's or unit's behaviors affect the behaviors of other members or units, there is an increased sense of interdependency. Interdependence leads to conflict. Team members in complex adaptive systems reflect the interdependence that exists in the larger organization. When people are dependent on each other the chance for conflict is significantly increased. Thus, conflict is an inevitable part of being connected.

Third, small changes in initial conditions can have large consequences. When teams are formed for the purposes of sensemaking, the fact that team members are connected to many other organizational members suggests that even subtle new insights achieved by a team may lead to large changes within the larger organization. Teams can transfer and nurture meaning, as well as amplify information inside the system in unexpected ways. The multiple perceptions, judgments, and interpretations afforded by team members set loose in a system where nonlinear outcomes are assumed to be true allow for the possibility that a small idea can eventually develop large consequences. It is the system's interconnectedness that makes such unexpected amplifications possible. Thus, understandings that get worked out in teams, or understandings yet to be worked out in teams, will find their way back to the larger organization. It should not be surprising that sometimes this amplifying effect will produce unexpected and new conflicts that will then need to be addressed by the organization.

For example, cross-functional teams commonly used to implementing total quality management are often charged with examining processes in order to identify and remove barriers to quality. Uncovering one barrier, and improving one aspect of a particular process can lead to significant quality enhancement that benefits the organization as a whole. Moreover, those members of the team may take back to their work units the experience of analyzing a process, defining quality, and creating appropriate measurements. Thus, the local work unit might learn from that original team experience and make similar improvements. What may seem like a small change to quality can become amplified as members of the entire organization are affected by the learning that took place among the original members of the cross-functional team.

Finally, teams that are authorized to experiment and innovate will create internal organizational instability. When teams explore new ideas, search for new ways to understand and do things, they push the organization out of its planned course, out of its stable equilibrium. Teams that are actively engaged in sensemaking and are allowed the freedom to self-organize are potentially a source of disruption and disorder. When pushing the organization away from cherished behavior and belief patterns, teams will undoubtedly create conflict for others in the organization.

Guidelines For Managing Conflict

Recognizing the pervasiveness of conflict in complex adaptive systems, managers should seek multiple frames, get comfortable with storming, and learn to actively surface conflict.

Seek Out Multiple Frames

In traditional organizations, conflict is seen as a problem that interferes with the accomplishment of purposes. In complex organizations it is useful for managers to think about conflict as evidence of another person's perceptual frame rather than as a problem that has to be eliminated. Teams that can learn to use conflict as an opportunity to see from another's perspective will be teams that are most able to help the organization make sense of its world. Bolman and Deal (1997) make this point in describing an exchange between the impressionist painter, Cezanne, and a critic. "A critic once commented to Cezanne, 'That doesn't look anything like a sunset.' Pondering his painting, Cezanne responded, 'Then you don't see sunsets the way I do'" (Bolman & Deal, 1997, p. 12).

In most organizations, an exchange such as the one above would result in the need to decide which view of the sunset is right. An alternative way of thinking of the conflict revealed in this exchange is to think of it as evidence of two different perceptual frames for thinking about the concept of sunset. A frame is both a window on the world and a lens that brings the world into focus. Organizations can experience the same clarity that Galileo discovered when he invented the telescope (Bolman & Deal, 1997). Each lens he added to the telescope contributed to a clearer image of the heavens. Teams can learn to do the same. Rather than resisting conflict, teams can use conflict to help them frame and reframe until they develop a collective understanding of the situation at hand. As Galileo added lenses for greater clarity, teams can use conflict to see clearer, see more, and see farther.

Get Comfortable with the Storming Stage

Traditional views of organizations and group behavior suggest that all groups go through four stages of group development: forming, storming, norming, and performing (Tuckman, 1965). In the forming stage, team members get oriented to the task and get acquainted with each other. The forming stage is followed by the storming stage in which individual personalities are expressed. During this stage, conflict and disagreement surface as members recognize that their perceptions, interests, and values are not aligned. In successful groups, the storming stage is followed by the norming stage in which conflicts are resolved and harmony and unity emerge. Finally, successful teams move from the norming stage into performing where task accomplishment is achieved. Most prescriptions for group development include the urgency of getting beyond the storming stage. For example, Daft and Marcic (1995, p. 519) warn, "unless teams can successfully move beyond this stage, they may get bogged down and never achieve high performance."

In contrast, the view of organizations as complex adaptive systems where teams have a responsibility for sensemaking suggests that the storming stage may in fact be the stage where teams spend most of their time. Rather than see this stage as something bad and to be avoided, teams need to learn to see the surfacing of conflict, the differing perspectives, the multiple frames as opportunities to develop the best understandings of the organization and its world.

Learn to Actively Surface Conflict

Although conflict surfacing techniques are not new (cf. Schwenk & Cosier, 1980), they generally receive much less attention in textbooks on organizational behavior and handbooks on group development than do conflict resolution discussions. Because people have such a strong aversion to conflict and because people's affiliative needs for inclusion and acceptance may be high in the early stages of using teams for sensemaking, it may be necessary for team leaders to actively surface the conflict that is there. Two methods for surfacing assumptions are devil's advocacy and dialectical inquiry (Schweiger & Sandberg, 1989; Schweiger, Sandberg, & Ragan, 1986). Both of these approaches offer the opportunity for groups to actually stimulate or surface conflict in an organized fashion.

The devil's advocacy approach develops a solid argument for a reasonable recommendation or decision. The recommendation is then subjected to an in-depth, formal critique which calls into question both the assumptions and recommendation. This in-depth critique (playing the role of the "devil's advocate") makes every attempt to show why the recommendation should not be adopted. Through repeated criticism and revision the approach is intended to lead to mutual acceptance of a recommendation. The approach is based on the assumption that good recommendations and assumptions will survive even the most forceful criticism (cf. Schweiger, Sandberg, & Ragan, 1986).

The dialectical inquiry approach is derived from Hegel's dialectic in which a high-quality decision is believed to result from pitting diametrically opposed assumptions and recommendations—a thesis and antithesis—against each other. Contrary recommendations and assumptions are developed from the same data, and are subject to critical evaluation through debate between advocate subgroups. The debaters attempt to spell out the implications of their recommendations, reveal the underlying assumptions, and challenge (or defend) those assumptions as effectively as possible. Following the debate, the subgroups are to decide on the assumptions which have survived the scrutiny of the debate and make a recommendation based on the surviving assumptions (Schweiger, Sandberg, & Ragan, 1986).

Note, that if managed effectively, neither of these conflict surfacing techniques need threaten the team's functioning. The team leader (or the person leading the conflict surfacing effort) needs to remind team members that they are playing *roles*, and not necessarily representing their personal points of view. The disassociation from oneself can be further emphasized by randomly assigning the roles. The team leader can also remind people that the team's goal is an effective decision, and reaching that goal requires a careful and systematic examination of ideas. Emphasizing the examination of ideas, and downplaying personalities can defuse potential threats to the team's social fabric.

SENSEMAKING TEAMS AND CONNECTEDNESS: TOWARD DEVELOPED MIND

Most prescriptions for successful teams stress the importance of team cohesiveness, defined as the extent to which members are attracted to the team and motivated to remain in it. From a traditional view of organizations, cohesiveness is developed through a greater amount of contact and interaction among group members, through agreement on goals and through personal attraction (i.e., similarity of attitudes and values) that leads to an enjoyment of being together (Cartwright & Zander, 1968).

Cohesive groups where intimate disclosure occurs are considered highly developed teams (Weick & Roberts, 1993). When teams have a primary focus on problem solving and task accomplishment, the development of intimacy allows group members the capacity to disclose information and views that might not be well received in other places but that are needed to address the problem at hand. In such groups, members share values, experience openness, and are willing to freely disclose. The more highly developed a team becomes, the more cohesiveness that is experienced by the team, the more delineated are the boundaries of the team. Teams that are highly developed as groups are often ones with impermeable boundaries, partly created by tight cohesion within the group (cf. Burton, 1990). That is, there is a clear sense of who is part of the team and who is not.

Enhancing cohesiveness (i.e., encouraging developed groups) will no doubt continue as an objective for teams in complex organizations, but as we have argued here, teams will also have another responsibility—developing a collective sense of mind (Weick & Roberts, 1993). As teams are used for a greater variety of purposes in organizations and as membership in teams is likely to be more and more fluid, attention to the creation of developed mind becomes critical. In particular, if teams are to have responsibility for helping an organization make sense of its world and collectively give meaning to its unfolding future, it is important to consider the emergence of the team as a collective mind, as well as a developed group.

Weick and Roberts (1993) describe collective mind as a pattern of heedful interrelations of actions in a social system. When groups act heedfully (i.e., as a collective mind), they act as if there were social forces pushing them to act, while envisioning a social system of joint action, and they construct their actions within that system. Weick and Roberts (1993) point out, however, that a team may be in a mature development stage as a group and yet undeveloped as a collective mind. When groupthink (Janis, 1982) occurs, it can be said that the team is a developed group (cohesive)—but acts heedlessly, an indication of undeveloped, collective mind. Similarly, it is possible for a team to be undeveloped as a group, yet act heedfully, as is the case with jazz improvisation. A highly developed collective mind emphasizes coordinating actions versus aligning cognitions, mutual respect versus agreement, trust versus empathy, and diversity versus homogeneity (Weick & Roberts, 1993).

The notion of collective mind underscores the importance of group members being connected without emphasizing the traditional role of cohesiveness. Being "connected" is not the same thing as being "cohesive." Being cohesive is about making everyone on the team see things the same way. Being connected is about heedful interrelating, which on the one hand enhances collective understanding, but on the other hand allows for diversity in self-identities. Connections are an important element of complex systems. As Kaufman (1995) points out, it is the connections among the agents that define the future state of a system. Effective teams in complex adaptive systems will be ones in which there is not so much an emphasis on cohesion as an acknowledgment of the connections that exist within the team and between the team and the organization. While traditional approaches to team building have focused on highly developed teams with impermeable boundaries and significant internal cohesiveness, the new model suggests a different emphasis—on developed mind.

Guidelines For Managing Connectedness

Managing connectedness requires that managers make intentional efforts to use teams so that interorganizational connections can be maximized. Enhancing connectedness leads to the emergence of developed mind. To facilitate developed mind, organizations should seek diversity in creating teams, and develop respect and trust versus agreement.

Insist on Diversity

Because the world is complicated, because information is available beyond most people's ability to retrieve or understand it, and because the environments in which many organizations exist experience rapid and discontinuous change, organizations as complex adaptive systems may be required to develop complicated patterns of relationships, if they are to make sense out of their worlds. For example, Ashby's Law of Requisite Variety (1956) suggests that the only way to cope with variety is with variety. Thus organizations existing in complex environments must match that external variety with internal variety. Stacey (1996) also notes that one of the factors that drives both the behavior and complexity of a system is the level or degree of diversity within and between the agents in the system. Thus, teams which are units created to help make sense out of a complex world will be effective to the extent that they develop for themselves complex relationships among their members, with the organization as a whole, and with the environment. Such teams will be capturing and replicating the complex relationships that exist in the organization and which are necessary for organizational functioning. Complex relationships include a diverse set of players who relate to each other and to others within the organization in a variety of ways. Such diversity and complexity will make it difficult and ultimately unnecessary for teams to experience the kind of cohesion that often is sought in teams in more traditional organization practices.

Traditional approaches to working with teams have emphasized the importance of developing similar values and similar goals (i.e., cohesiveness). Teams in complex adaptive systems are likely to be more successful if they mirror the complexity that is present elsewhere in the organization. Heterogeneous teams made up of individuals loosely coupled to many other parts of the system (and outside the system) can best contribute to the conditions for developing a rich, collective mind.

Develop Trust versus Agreement

The development of collective mind is more dependent on mutual respect and trust than on agreement and acceptance. When interactions among team members are respectful, people do three things, according to Weick (1996): (1) they respect the reports and assertions of others, and are willing to base their beliefs and actions on those reports; (2) they faithfully and accurately report so that others may use their observations in coming to new beliefs; and (3) they respect their own perceptions and beliefs and integrate them with the reports of others. These three activities represent a self aware and thoughtful system's features of trust, honesty, and self-respect.

SENSEMAKING TEAMS AND LEADERSHIP

As the charge of teams in complex adaptive organizations expands to include the process of sensemaking, the nature of leadership in teams requires reexamination.

Traditional approaches to management view leadership as the ability to influence people to achieve some end or goal (Daft & Marcic, 1995). This definition implies a tight connection between cause and effect and is grounded in the more machine-like image of organizations in which people do what others push them to do, where a priori goals are set and leaders encourage people to achieve those goals. And so it has been with teams. Effective team leaders, from a traditional view are the captains who are able to articulate a clear vision and set of goals, who encourage commitment to the goals, and who remove obstacles that get in the team's way of achieving those goals (Katzenbach & Smith, 1993). But this view of leadership does not fit well with the perspective of organizations as complex adaptive systems, operating in a complex world facing unknowable futures.

> Only under familiar conditions can the captain identify the ship's future destination, and only under such conditions does it make sense for members of the team to follow the leader slavishly. An old map is useless when the terrain is new. Old beliefs cannot help in the task managers face today: managing the unknowable. (Stacey, 1992, p. 4)

If teams are to be used for helping the organization make sense of its world, leaders of sensemaking teams will be required to function differently. As Wheatley (1997, p. 25) points out, "most of us were raised in a culture that told us that the way to manage for excellence was to tell people exactly what they had to do and then make sure they did it." The assumption of the old model of leadership was that the leader knew what needed to be done, and if enough plans were made people could be engineered into doing what the master engineer had in mind.

Leadership Guidelines

In complex adaptive systems a new approach to leadership will be required. It is an approach in which team leaders will support self-organizing responses, be agents of disturbance instead of agents of alignment, and act as resource providers.

Support Self-Organizing Responses

Teams that have a sensemaking responsibility in complex adaptive systems will be self-organizing units. As teams work toward developing collective understanding of what is happening to the organization and its world, they will develop ideas, and possibly spontaneously reorganize as they envision new issues to be addressed and identify alternative ways of addressing them. Patterns of understanding and communication will change as members of teams come and go depending on the evolving focus of the team. Leaders of such teams need to be capable of letting understanding and actions unfold and finding ways to support the new patterns that emerge. The leader's job is to encourage creativity and innovation rather than to give guidelines and monitor group behavior to see that it stays within the guidelines. Ghoshal and Barlett (1995) describe this philosophy at 3M, a company known for emergent ideas, by quoting the company's founder, William L. McKnight:

> Mistakes will be made, but if a person is essentially right, the mistakes he or she makes are not nearly as serious in the long run as the mistakes management will make if it is dictatorial and undertakes to tell those under its authority how they must do their jobs (p. 89).

Part of the way self-organizing responses are supported by the leader is to encourage self-leadership. To promote innovation and creativity, complex adaptive systems need more than just a small number of leaders who do the thinking for others, but rather numerous capable leaders working through the organization (Senge, 1990). Team leaders must encourage individuals towards self-leadership, a process in which everyone participates. Within teams, each team member assumes responsibility for the group's results and the group leader is the person who happens to assume the responsibility agreed to by the group itself. Team members will need to foster individual leadership rather than individual obedience.

Be Agents of Disturbance Instead of Agents of Alignment

One of the tenets of a complex adaptive systems view of organizations is that it is possible (even desirable) that systems exist in a state far from equilibrium. Systems experiment and explore new possibilities that result in new patterns and structures and new sets of relationships. Similarly, teams within complex adaptive systems do the same. In fact, sensemaking teams may be the disturbing agents that push the organization out of its state of equilibrium and into some new place. Leaders of sensemaking teams should see their role as encouraging disturbance and resisting alignment with past practices and past goals. Leaders of teams legitimate the questioning of corporate dogma and assumptions inherent in it because such questioning must first happen before growth and renewal can take place. That is, current assumptions must be challenged and overturned before new, more adaptive patterns can emerge. As agents of disturbance, team leaders enable disturbance to create choice points (bifurcation points) within the team which can lead to new forms of self-organizing. As agents of disturbance, team leaders must resist the temptation to influence team members to align with each other or the leader.

Provide Resources Needed by the Team

Senge (1990) notes that traditional views of leaders—as special people who set the direction, make the key decisions, and energize the troops—are deeply rooted in an individualistic and nonsystemic worldview. Rather than be the captains, leaders in self-organizing sensemaking teams are more the "designers" that Senge describes. When teams face unknowable worlds and seek understandings that reduce the equivocality they face, when ideas and patterns emerge and teams self-organize, the leader's role is to help create an environment where creativity and emergent ideas can occur. Rather than provide teams with goals and solutions, leaders in the new organization need to provide teams with resources and tools that make self-initiatives possible.

The role of leaders, as described by Ghoshal and Bartlett (1995, p. 96) is to "create an environment that enables [people] to take initiative, to cooperate and to learn." These leaders will foster experimentation and help create connections across the organization in order to feed the team with the rich information from multiple sources.

CONCLUSION

We have argued that viewing organizations through the conceptual lens of complex systems brings into focus insights about teams that would go unnoticed through a different, more traditional conceptual lens. A complex systems view of organizations places less emphasis on command and control and instead focuses on the emergent properties of the system as it tries to adapt to environmental change. The complex systems view also emphasizes connectedness, the network of interconnections inside the system. Because of the assumptions built into the complex adaptive systems approach, organizations must cope with an unknowable future, and teams will play an important, perhaps even essential, role in helping the organization make sense of its world. This sensemaking role will come in addition to a team's traditional roles of information processing and task accomplishment and will change our notions of what constitutes appropriate team and leader behavior.

Three important dimensions of team behavior will be approached differently as a consequence of applying a complex adaptive systems perspective to organizations: conflict, connectedness, and leadership. Sensemaking teams will be expected to experience conflict and, rather than try to eliminate it, will use conflict as a mechanism for achieving new insights and interpretations. Sensemaking teams will consciously seek out differing perspectives and multiple frames. Moreover, sensemaking teams will have to become skilled at negotiating the "storming stage" of group behavior in order to surface conflict. By seeking multiple frames and differing points of view, sensemaking teams will bypass an effort to achieve cohesiveness and will instead focus on developing a collective sense of purpose. A collective sense of purpose for a team will mean that the team develops a degree of respect and trust that will allow the team to make sense of the world without insisting that each team member reach the kind of uniform identity which has traditionally been the hallmark of "good" groups. Finally, effective leadership in sensemaking teams will also be different. Leaders of successful sensemaking teams will encourage the team's self-organizing capabilities. They will be agents of disturbance, not agents of alignment, and they will direct their efforts at providing the team its needed resources.

Using complex adaptive systems as the basis for understanding organizations will require considerable changes to the way we think about the practice of management. Most notable of these changes is the idea that it will be necessary to intentionally create "disorder" so that new, adaptive patterns can emerge. But a complex, turbulent world will require complex, turbulent responses. Letting go of the belief that safe, predictable actions are possible may be the greatest challenge managers face in the new world of management.

REFERENCES

Ashby, W.R. (1956). *An introduction to cybernetics.* London: Chapman and Hall.

Ashmos, D.P. (1997). Building effective health care teams. In W.J. Duncan, P. Gunter, & L. Swain (Eds.), *Handbook of health care management* (pp. 313-338). Oxford, England: Blackwell Press.

Ashmos, D.P., & Nathan, M.L. (1998). Why (and how) sensemaking teams make sense. Unpublished paper, University of Texas at San Antonio.

Bohm, D. (1983). *Wholeness and the implicate order.* Boston: Ark 21.

Bolman, L.G., & Deal, T.E. (1997). *Reframing organizations.* San Francisco: Jossey Bass Publishers.

Burrell, G., & Morgan, G. (1979). *Sociological paradigms and organizational analysis.* London: Heineman.

Burton, G.E. (1990, February). The measurement of distortion tendencies induced by the win-lose nature of in-group loyalty. *Small Group Research*, 128-141.

Capra, F. (1982). *The turning point: Science, society, and the rising culture.* London: Flamingo.

Capra, F. (1996). *The web of life.* New York: Doubleday.

Cartwright, C., & Zander, A. (1968). *Group dynamics: Research and theory.* New York: Harper & Row.

Cohen, J., & Stewart, I. (1994). *The collapse of chaos.* New York: Viking Press.

Daft, R.L., & Marcic, D. (1995). *Understanding management.* Ft. Worth, TX: Dryden Press.

Gagne, T.E. (1995). *Designing effective organizations: Traditional and transformational views.* Thousand Oaks, CA: Sage Publications.

Ghoshal, S. & Bartlett, C.A. (1995, January-February). Changing the role of top management. *Harvard Business Review*, 86-96.

Herbert, Nick. (1985). *Quantum reality: Beyond the new physics.* New York: Doubleday.

Janis, I.L. (1982). *Groupthink: Psychological studies of policy decisions and fiascoes.* Boston: Houghton Mifflin.

Katzenbach, J.R., & Smith, D.K. (1993). *The wisdom of teams.* New York: Harper Business.

Kaufman, S. (1995). *At home in the universe.* Oxford, England: Oxford University Press.

Locke, E., & Schweiger, D. (1979). Participation in decision making: A review. In Staw, B. (Ed.), *Research in organizational behavior* (vol. 1, pp. 265-339). Greenwich, CT: JAI Press, Inc.

March, J., & Simon, H. (1993). *Organizations, 2nd edition.* Oxford, England: Blackwell Press.

Maslow, A. (1965). *Eupsychian management.* Homewood, IL: Irwin.

McDaniel, R.R., Jr. (1997). Strategic leadership: A view from quantum and chaos theories. In W.J. Duncan, P. Ginter, L. Swayne, (Eds.), *Handbook of healthcare management* (pp. 339-367). Oxford, England: Blackwell Press.

Overman, E.S. (1996). The new science of management: Chaos and quantum theory and method. *Journal of Public Administration Research and Theory, 6,* 75-89.

Prigogine, I., & Stengers, I. (1984). *Order out of chaos: Man's new dialogue with nature.* New York: Bantam Books.

Rae, A. (1986). *Quantum physics: Illusion or reality?* Cambridge, MA: Cambridge University Press.

Robbins, H., & Finley, H. (1995). *Why teams don't work.* Princeton, NJ: Peterson's/Pacesetter Books.

Rohrlich, F. (1987). *From paradox to reality: Our basic concepts of the physical world.* Cambridge, MA: Cambridge University Press.

Schweiger, D.M., & Sandberg, W.R. (1989). The utilization of individual capabilities in group approaches to strategic decision making. *Strategic Management Journal, 10,* 31-43.

Schweiger, D.M., Sandberg, W.R., & Ragan, J.W. (1986). Group approaches for improving strategic decision making: A comparative analysis of dialectical inquiry, devil's advocacy, and consensus. *Academy of Management Journal, 29,* 51-71.

Schwenk, C.R., & Cosier, R.A. (1980). Effects of the expert, devil's advocate, and dialectical inquiry methods on prediction performance. *Organizational Behavior and Human Performance, 26,* 409-424.

Senge, P.M. (1990). *The fifth discipline.* New York: Doubleday/Currency.

Smircich, L., & Stubbart, C. (1985). Strategic management in an enacted world. *Academy of Management Review, 10*(4), 724-736.

Stacey, R.D. (1992). *Managing the unknowable: Strategic boundaries between order and chaos in organizations.* San Francisco: Jossey-Bass Publishers.

Stacey, R.D. (1995). The science of complexity: An alternative perspective for strategic change processes. *Strategic Management Journal, 16*(6), 477-495.

Stacey, R.D. (1996). *Complexity and creativity in organizations.* San Francisco: Berrett-Koehler.

Tjosvold, D., & Tjosvold, M.M. (1995). Cross-functional teamwork: The challenge of involving professionals. In M.M. Beyerlein, D.A. Johnson, & S.T. Beyerlein (Eds.), *Advances in interdisciplinary studies of work teams, Vol. 2: Knowledge work in teams* (pp. 1-34). Greenwich, CT: JAI Press.

Tuckman, B.W. (1965, November). Development sequence in small groups. *Psychological Bulletin,* 384-399.

Weick, K.E. (1993). The collapse of sensemaking in organizations: The mann gulch disaster. *Administrative Science Quarterly, 38,* 628-652.

Weick, K.E. (1995). *Sensemaking in organizations.* Thousand Oaks, CA: Sage Publications, Inc.

Weick, K. (1996, May-June). Prepare your organizations to fight fire. *Harvard Business Review,* 145-148.

Weick, K.E. & Roberts, K.H. (1993). Collective mind in organizations: Heedful interrelating on flight decks. *Administrative Science Quarterly, 38*(3), 357-381.

Wheatley, M.J. (1992). *Leadership and the new science: Learning about organization from an orderly universe.* San Francisco Berrett-Koehler Publishers, Inc.

Wheatley, M.J. (1997, Summer). Goodbye, command and control. *Leader to Leader,* 21-28.

Youngblood, M.D. (1997) *Life at the edge of chaos: Creating the quantum organization.* Dallas, TX: Perceval Publishing.

ABOUT THE EDITORS

Michael M. Beyerlein is director of the Center for the Study of Work Teams and associate professor of psychology at the University of North Texas. His research interests include all aspects of work teams, organizational transformation, job stress, creativity/innovation, knowledge management, the learning organization, and complex adaptive systems. He has published in a number of research journals, is a member of the editorial boards for *TEAM Magazine* and *Quality Management Journal*, and edits *Infrastructure: Sustaining Systems*, an e-journal. In addition, he has been a co-editor on two case books about teams, has an edited book on the history of work teams in press, and is currently co-editing a book on sustaining teams.

Douglas A. Johnson is director of the Industrial/Organizational psychology program, professor of psychology, and associate director of the Center for the Study of Work Teams, University of North Texas. Doug has published research in a variety of areas, ranging from leadership and job satisfaction to operant conditioning and interpersonal attraction. He co-founded and served as president of the Dallas-Fort Worth Organizational Psychology Group, and participated in the creation of the Dallas office of the I/O psychology consulting firm, Personnel Decisions International, with whom he works on a part-time basis.

Susan T. Beyerlein has taught undergraduate and MBA management courses at Our Lady of the Lake University and Texas Woman's University in the Dallas area. Susan has served as a research project manager with the Center for the Study of Work Teams and as a research scientist with the Center for Public Management at the University of North Texas. She continues to be an ad hoc reviewer for the *Academy of Management Review*. She is currently working on several edited book projects.